The Mathematical Theory of Communication

Dear Annie,

Congratulations on your graduation and starting grad school.

While I won't describe this book as a "fun" read, it's certainly one that should be on the bookshelf of every researcher in the field.

It has been a pleasure working with you and I'm proud of what you've accomplished and wish you the best for the future.

Siddhartan
siddgov@alum.mit.edu

4/30/2010

THE MATHEMATICAL THEORY
OF COMMUNICATION

CLAUDE E. SHANNON

and

WARREN WEAVER

Foreword by
Richard E. Blahut and Bruce Hajek

University of Illinois Press

Urbana and Chicago

Contents

Foreword

The story is told that young King Solomon was given
the choice between wealth and wisdom. When he chose
wisdom, God was so pleased that he gave Solomon not only
wisdom but wealth also. So it is with science.

—Arthur Holly Compton

A half century ago, in 1948, Claude Shannon published his epic
paper "A Mathematical Theory of Communication." This paper
was thereafter republished in book form in 1949 by the University
of Illinois Press, together with an expository introduction by
Warren Weaver. The slight change in the title between 1948 and
1949 is both trivial and profound. The paper and, by extension,
the book have had an immense impact on technological progress,
and so on life as we now know it. Despite its specialized character, the book has sold more than 40,000 copies and continues to
sell 700 or more copies per year. It is required reading in many
courses on information theory. We are delighted to have a small
role in the publication of this golden anniversary edition, which is
identical to the original except for the correction of a number of
minor typographical errors that have existed through four hard
cover and sixteen paperback printings.

In this important work, Shannon chose to use a light touch and
a gentle delivery more common in technical papers of that time
than in our own. He saw deeply into the essence of the communication problem and chose to deliver that wisdom in his own way
and with a minimum of mathematical proof. Perhaps this under-
lies the timeless quality of the work. The proofs came elsewhere
and later, but the insight shines through. One measure of the
greatness of the book is that Shannon's major precept that all
communication is essentially digital is now commonplace among

the modern digitalia, even to the point where many wonder why Shannon needed to state such an obvious axiom. Yet his audience fifty years ago took a somewhat skeptical and aloof view of his work.

Shannon had the presight to overlay the subject of communication with a distinct partitioning into *sources, source encoders, channel encoders, channels,* and *associated channel and source decoders.* Although his formalization seems quite obvious in our time, it was not so obvious back then. Shannon further saw that channels and sources could and should be described using the notions of entropy and conditional entropy. He argued persuasively for the use of these notions, both through their characterization by intuitive axioms and by presentation of precise coding theorems. Moreover, he indicated how very explicit, operationally significant concepts such as the information content of a source or the information capacity of a channel can be identified using entropy and maximization of functions involving entropy.

Shannon's revolutionary work brought forth this new subject of information theory fully formed but waiting for the maturity that fifty years of aging would bring. It is hard to imagine how the subject could have been created in an evolutionary way, though after the conception its evolution proceeded in the hands of hundreds of authors to produce the subject in its current state of maturity.

The exposition by Warren Weaver that introduces the book is one of his many and diverse contributions toward promoting the understanding of science and mathematics to a broad audience. It illustrates how Shannon's ideas have implications that were (at least fifty years ago) well beyond the immediate goals of communication engineers and of Shannon himself. These include insights for linguists and for social scientists addressing broad communication issues.

The impact of Shannon's theory of information on the development of telecommunication has been immense. This is evident to those working at the edge of advancing developments, though perhaps not quite so visible to those involved in routine design. The notion that a channel has a specific information capacity, which can be measured in bits per second, has had a profound influence. On the one hand, this notion offers the promise, at least in theory, of communication systems with frequency of errors as small as desired for a given channel for any data rate less than the channel capacity. Moreover, Shannon's associated existence proof provided tantalizing insight into how ideal communication systems might someday fulfill the promise. On the other hand, this notion also clearly establishes a limit on the communication rate that can be achieved over a channel, offering communication engineers the ultimate benchmark with which to calibrate progress toward construction of the ultimate communication system for a given channel.

The fact that a specific capacity can be reached, and that no data transmission system can exceed this capacity, has been the holy grail of modem design for the last fifty years. Without the guidance of Shannon's capacity formula, modem designers would have stumbled more often and proceeded more slowly. Communication systems ranging from deep-space satellite links to storage devices such as magnetic tapes and ubiquitous compact disks, and from high-speed internets to broadcast high-definition television, came sooner and in better form because of his work. Aside from this wealth of consequences, the wisdom of Claude Shannon's insights may in the end be his greatest legacy.

Richard E. Blahut
Bruce Hajek

Preface

Recent years have witnessed considerable research activity in communication theory by a number of workers both here and abroad. In view of the widespread interest in this field, Dean L. N. Ridenour suggested the present volume consisting of two papers on this subject.

The first paper has not previously been printed in its present form, although a condensation appeared in *Scientific American*, July, 1949. In part, it consists of an expository introduction to the panoramic view of the field before entering into the more mathematical aspects. In addition, some ideas are suggested for broader application of the fundamental principles of communication theory.

The second paper is reprinted from the *Bell System Technical Journal*, July and October, 1948, with no changes except the correction of minor errata and the inclusion of some additional references. It is intended that subsequent developments in the field will be treated in a projected work dealing with more general aspects of information theory.

It gives us pleasure to express our thanks to Dean Ridenour for making this book possible and to the University of Illinois Press for their splendid cooperation.

C. E. Shannon
W. Weaver

RECENT CONTRIBUTIONS TO THE MATHEMATICAL THEORY OF COMMUNICATION

By Warren Weaver

1
Introductory Note on the General Setting of the Analytical Communication Studies[1]

1.1. Communication

The word *communication* will be used here in a very broad sense to include all of the procedures by which one mind may affect another. This, of course, involves not only written and oral speech, but also music, the pictorial arts, the theatre, the ballet, and in fact all human behavior. In some connections it may be desirable to use a still broader definition of communication, namely, one which would include the procedures by means of which one mechanism (say automatic equipment to track an airplane and to compute its probable future positions) affects another mechanism (say a guided missile chasing this airplane).

The language of this memorandum will often appear to refer to the special, but still very broad and important, field of the communication of speech; but practically everything said applies

[1] This paper is written in three main sections. In the first and third, W. W. is responsible both for the ideas and the form. The middle section, namely "2), Communication Problems of Level A" is an interpretation of mathematical papers by Dr. Claude E. Shannon of the Bell Telephone Laboratories. Dr. Shannon's work roots back, as von Neumann has pointed out, to Boltzmann's observation, in some of his work on statistical physics (1894), that entropy is related to "missing information," inasmuch as it is related to the number of alternatives which remain possible to a physical system after all the macroscopically observable information concerning it has been recorded. L. Szilard (Zsch. f. Phys. Vol. 53, 1925) extended this idea to a general discussion of information in physics, and von Neumann (*Math. Foundation of Quantum Mechanics,* Berlin, 1932, Chap. V) treated information in quantum mechanics and particle physics. Dr. Shannon's work connects more directly with certain ideas developed some twenty years ago by H. Nyquist and R. V. L. Hartley, both of the Bell Laboratories; and Dr. Shannon has himself emphasized that communication theory owes a great debt to Professor Norbert Wiener for much of its basic philosophy. Professor Wiener, on the other hand, points out that Shannon's early work on switching and mathematical logic antedated his own interest in this field; and generously adds that Shannon certainly deserves credit for independent development of such fundamental aspects of the theory as the introduction of entropic ideas. Shannon has naturally been specially concerned to push the applications to engineering communication, while Wiener has been more concerned with biological application (central nervous system phenomena, etc.).

equally well to music of any sort, and to still or moving pictures, as in television.

1.2. Three Levels of Communications Problems

Relative to the broad subject of communication, there seem to be problems at three levels. Thus it seems reasonable to ask, serially:

LEVEL A. How accurately can the symbols of communication be transmitted? (The technical problem.)

LEVEL B. How precisely do the transmitted symbols convey the desired meaning? (The semantic problem.)

LEVEL C. How effectively does the received meaning affect conduct in the desired way? (The effectiveness problem.)

The *technical problems* are concerned with the accuracy of transference from sender to receiver of sets of symbols (written speech), or of a continuously varying signal (telephonic or radio transmission of voice or music), or of a continuously varying two-dimensional pattern (television), etc. Mathematically, the first involves transmission of a finite set of discrete symbols, the second the transmission of one continuous function of time, and the third the transmission of many continuous functions of time or of one continuous function of time and of two space coordinates.

The *semantic problems* are concerned with the identity, or satisfactorily close approximation, in the interpretation of meaning by the receiver, as compared with the intended meaning of the sender. This is a very deep and involved situation, even when one deals only with the relatively simpler problems of communicating through speech.

One essential complication is illustrated by the remark that if Mr. X is suspected not to understand what Mr. Y says, then it is theoretically not possible, by having Mr. Y do nothing but talk further with Mr. X, completely to clarify this situation in any finite time. If Mr. Y says "Do you now understand me?" and Mr. X says "Certainly, I do," this is not necessarily a certification that understanding has been achieved. It may just be that Mr. X did not understand the question. If this sounds silly, try it again as "Czy pañ mnie rozumie?" with the answer "Hai wakkate imasu." I think that this basic difficulty[2] is, at least in the restricted field of speech communication, reduced to a tolerable size (but never completely eliminated) by "explanations" which (a) are presumably never more than approximations to the ideas being explained, but which (b) are understandable since they are phrased in language which has previously been made reasonably clear by operational means. For example, it does not take long to make the symbol for "yes" in any language operationally understandable.

The semantic problem has wide ramifications if one thinks of communication in general. Consider, for example, the meaning to a Russian of a U.S. newsreel picture.

The *effectiveness problems* are concerned with the success with which the meaning conveyed to the receiver leads to the desired conduct on his part. It may seem at first glance undesirably narrow to imply that the purpose of all communication is to influence the conduct of the receiver. But with any reasonably broad definition of conduct, it is clear that communication either affects conduct or is without any discernible and probable effect at all.

The problem of effectiveness involves aesthetic considerations in the case of the fine arts. In the case of speech, written or oral, it involves considerations which range all the way from the mere mechanics of style, through all the psychological and emotional aspects of propaganda theory, to those value judgments which are necessary to give useful meaning to the words "success" and "desired" in the opening sentence of this section on effectiveness.

The effectiveness problem is closely interrelated with the semantic problem, and overlaps it in a rather vague way; and

[2] "When Pfungst (1911) demonstrated that the horses of Elberfeld, who were showing marvelous linguistic and mathematical ability, were merely reacting to movements of the trainer's head, Mr. Krall (1911), their owner, met the criticism in the most direct manner. He asked the horses whether they could see such small movements and in answer they spelled out an emphatic 'No.' Unfortunately we cannot all be so sure that our questions are understood or obtain such clear answers." See Lashley, K. S., "Persistent Problems in the Evolution of Mind" in *Quarterly Review of Biology*, v. 24, March, 1949, p. 28.

there is in fact overlap between all of the suggested categories of problems.

1.3. Comments

So stated, one would be inclined to think that Level A is a relatively superficial one, involving only the engineering details of good design of a communication system; while B and C seem to contain most if not all of the philosophical content of the general problem of communication.

The mathematical theory of the engineering aspects of communication, as developed chiefly by Claude Shannon at the Bell Telephone Laboratories, admittedly applies in the first instance only to problem A, namely, the technical problem of accuracy of transference of various types of signals from sender to receiver. But the theory has, I think, a deep significance which proves that the preceding paragraph is seriously inaccurate. Part of the significance of the new theory comes from the fact that levels B and C, above, can make use only of those signal accuracies which turn out to be possible when analyzed at Level A. Thus any limitations discovered in the theory at Level A necessarily apply to levels B and C. But a larger part of the significance comes from the fact that the analysis at Level A discloses that this level overlaps the other levels more than one could possible naively suspect. Thus the theory of Level A is, at least to a significant degree, also a theory of levels B and C. I hope that the succeeding parts of this memorandum will illuminate and justify these last remarks.

2
Communication Problems at Level A

2.1. A Communication System and Its Problems

The communication system considered may be symbolically represented as follows:

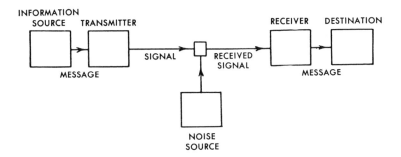

The *information source* selects a desired *message* out of a set of possible messages (this is a particularly important remark, which requires considerable explanation later). The selected message may consist of written or spoken words, or of pictures, music, etc.

The *transmitter* changes this *message* into the *signal* which is actually sent over the *communication channel* from the transmitter to the *receiver*. In the case of telephony, the channel is a wire, the signal a varying electrical current on this wire; the transmitter is the set of devices (telephone transmitter, etc.) which change the sound pressure of the voice into the varying electrical current. In telegraphy, the transmitter codes written words into sequences of interrupted currents of varying lengths (dots, dashes, spaces). In oral speech, the information source is the brain, the transmitter is the voice mechanism producing the varying sound pressure (the signal) which is transmitted through the air (the channel). In radio, the channel is simply space (or the aether, if any one still prefers that antiquated and misleading word), and the signal is the electromagnetic wave which is transmitted.

The *receiver* is a sort of inverse transmitter, changing the transmitted signal back into a message, and handing this message on to the destination. When I talk to you, my brain is the information source, yours the destination; my vocal system is the transmitter, and your ear and the associated eighth nerve is the receiver.

In the process of being transmitted, it is unfortunately characteristic that certain things are added to the signal which were not intended by the information source. These unwanted additions may be distortions of sound (in telephony, for example) or static (in radio), or distortions in shape or shading of picture (television), or errors in transmission (telegraphy or facsimile), etc. All of these changes in the transmitted signal are called *noise*.

The kind of questions which one seeks to ask concerning such a communication system are:

a. How does one measure *amount of information?*

b. How does one measure the *capacity* of a communication channel?

c. The action of the transmitter in changing the message into the signal often involves a *coding process*. What are the characteristics of an efficient coding process? And when the coding is as efficient as possible, at what rate can the channel convey information?

d. What are the general characteristics of *noise?* How does noise affect the accuracy of the message finally received at the destination? How can one minimize the undesirable effects of noise, and to what extent can they be eliminated?

e. If the signal being transmitted is *continuous* (as in oral speech or music) rather than being formed of *discrete* symbols (as in written speech, telegraphy, etc.), how does this fact affect the problem?

We will now state, without any proofs and with a minimum of mathematical terminology, the main results which Shannon has obtained.

2.2. Information

The word *information*, in this theory, is used in a special sense that must not be confused with its ordinary usage. In particular, *information* must not be confused with meaning.

In fact, two messages, one of which is heavily loaded with meaning and the other of which is pure nonsense, can be exactly equivalent, from the present viewpoint, as regards information. It is this, undoubtedly, that Shannon means when he says that "the semantic aspects of communication are irrelevant to the engineering aspects." But this does not mean that the engineering aspects are necessarily irrelevant to the semantic aspects.

To be sure, this word information in communication theory relates not so much to what you *do* say, as to what you *could* say.

That is, information is a measure of one's freedom of choice when one selects a message. If one is confronted with a very elementary situation where he has to choose one of two alternative messages, then it is arbitrarily said that the information, associated with this situation, is unity. Note that it is misleading (although often convenient) to say that one or the other message conveys unit information. The concept of information applies not to the individual messages (as the concept of meaning would), but rather to the situation as a whole, the unit information indicating that in this situation one has an amount of freedom of choice, in selecting a message, which it is convenient to regard as a standard or unit amount.

The two messages between which one must choose, in such a selection, can be anything one likes. One might be the text of the King James Version of the Bible, and the other might be "Yes." The transmitter might code these two messages so that "zero" is the signal for the first, and "one" the signal for the second; or so that a closed circuit (current flowing) is the signal for the first, and an open circuit (no current flowing) the signal for the second. Thus the two positions, closed and open, of a simple relay, might correspond to the two messages.

To be somewhat more definite, the amount of information is defined, in the simplest cases, to be measured by the logarithm of the number of available choices. It being convenient to use logarithms[3] to the base 2, rather than common or Briggs' logarithm to the base 10, the information, when there are only two choices, is proportional to the logarithm of 2 to the base 2. But this is unity; so that a two-choice situation is characterized by information of unity, as has already been stated above. This unit of information is called a "bit," this word, first suggested by John W. Tukey, being a condensation of "binary digit." When numbers are expressed in the binary system there are only two digits, namely 0 and 1; just as ten digits, 0 to 9 inclusive, are used in the decimal number system which employs 10 as a base. Zero and one may be taken symbolically to represent any two choices, as noted above; so that "binary digit" or "bit" is natural to associate with the two-choice situation which has unit information.

If one has available say 16 alternative messages among which

[3] When $m^x = y$, then x is said to be the logarithm of y to the base m.

he is equally free to choose, then since $16 = 2^4$ so that $\log_2 16 = 4$, one says that this situation is characterized by 4 bits of information.

It doubtless seems queer, when one first meets it, that information is defined as the *logarithm* of the number of choices. But in the unfolding of the theory, it becomes more and more obvious that logarithmic measures are in fact the natural ones. At the moment, only one indication of this will be given. It was mentioned above that one simple on-or-off relay, with its two positions labeled, say, 0 and 1 respectively, can handle a unit information situation, in which there are but two message choices. If one relay can handle unit information, how much can be handled by say three relays? It seems very reasonable to want to say that three relays could handle three times as much information as one. And this indeed is the way it works out if one uses the logarithmic definition of information. For three relays are capable of responding to 2^3 or 8 choices, which symbolically might be written as 000, 001, 011, 010, 100, 110, 101, 111, in the first of which all three relays are open, and in the last of which all three relays are closed. And the logarithm to the base 2 of 2^3 is 3, so that the logarithmic measure assigns three units of information to this situation, just as one would wish. Similarly, doubling the available time squares the number of possible messages, and doubles the logarithm; and hence doubles the information if it is measured logarithmically.

The remarks thus far relate to artificially simple situations where the information source is free to choose only between several definite messages — like a man picking out one of a set of standard birthday greeting telegrams. A more natural and more important situation is that in which the information source makes a sequence of choices from some set of elementary symbols, the selected sequence then forming the message. Thus a man may pick out one word after another, these individually selected words then adding up to form the message.

At this point an important consideration which has been in the background, so far, comes to the front for major attention. Namely, the role which probability plays in the generation of the message. For as the successive symbols are chosen, these choices are, at least from the point of view of the communication system, governed by probabilities; and in fact by probabilities which are not independent, but which, at any stage of the process, depend upon the preceding choices. Thus, if we are concerned with English speech, and if the last symbol chosen is "the," then the probability that the next word be an article, or a verb form other than a verbal, is very small. This probabilistic influence stretches over more than two words, in fact. After the three words "in the event" the probability for "that" as the next word is fairly high, and for "elephant" as the next word is very low.

That there are probabilities which exert a certain degree of control over the English language also becomes obvious if one thinks, for example, of the fact that in our language the dictionary contains no words whatsoever in which the initial letter j is followed by b, c, d, f, g, j, k, l, q, r, t, v, w, x, or z; so that the probability is actually zero that an initial j be followed by any of these letters. Similarly, anyone would agree that the probability is low for such a sequence of words as "Constantinople fishing nasty pink." Incidentally, it is low, but not zero; for it is perfectly possible to think of a passage in which one sentence closes with "Constantinople fishing," and the next begins with "Nasty pink." And we might observe in passing that the unlikely four-word sequence under discussion *has* occurred in a single good English sentence, namely the one above.

A system which produces a sequence of symbols (which may, of course, be letters or musical notes, say, rather than words) according to certain probabilities is called a *stochastic process,* and the special case of a stochastic process in which the probabilities depend on the previous events, is called a *Markoff process* or a Markoff chain. Of the Markoff processes which might conceivably generate messages, there is a special class which is of primary importance for communication theory, these being what are called *ergodic processes.* The analytical details here are complicated and the reasoning so deep and involved that it has taken some of the best efforts of the best mathematicians to create the associated theory; but the rough nature of an ergodic process is easy to understand. It is one which produces a sequence of symbols which would be a poll-taker's dream, because any reasonably large sample tends to be representative of the sequence as a whole. Suppose that two persons choose samples in different ways, and study what trends their statistical properties would show as the samples become larger. If the situation is ergodic, then those two persons, however they may have chosen their samples, agree in their estimates of the properties of the whole. Ergodic systems, in other words, exhibit a particularly safe and comforting sort of statistical regularity.

Now let us return to the idea of *information*. When we have an information source which is producing a message by successively selecting discrete symbols (letters, words, musical notes, spots of a certain size, etc.), the probability of choice of the various symbols at one stage of the process being dependent on the previous choices (i.e., a Markoff process), what about the information associated with this procedure?

The quantity which uniquely meets the natural requirements that one sets up for "information" turns out to be exactly that which is known in thermodynamics as *entropy*. It is expressed in terms of the various probabilities involved — those of getting to certain stages in the process of forming messages, and the probabilities that, when in those stages, certain symbols be chosen next. The formula, moreover, involves the *logarithm* of probabilities, so that it is a natural generalization of the logarithmic measure spoken of above in connection with simple cases.

To those who have studied the physical sciences, it is most significant that an entropy-like expression appears in the theory as a measure of information. Introduced by Clausius nearly one hundred years ago, closely associated with the name of Boltzmann, and given deep meaning by Gibbs in his classic work on statistical mechanics, entropy has become so basic and pervasive a concept that Eddington remarks "The law that entropy always increases — the second law of thermodynamics — holds, I think, the supreme position among the laws of Nature."

In the physical sciences, the entropy associated with a situation is a measure of the degree of randomness, or of "shuffledness" if you will, in the situation; and the tendency of physical systems to become less and less organized, to become more and more perfectly shuffled, is so basic that Eddington argues that it is primarily this tendency which gives time its arrow — which would reveal to us, for example, whether a movie of the physical world is being run forward or backward.

Thus when one meets the concept of entropy in communication theory, he has a right to be rather excited — a right to suspect that one has hold of something that may turn out to be basic and important. That information be measured by entropy is, after all, natural when we remember that information, in communication theory, is associated with the amount of freedom of choice we have in constructing messages. Thus for a communication source one can say, just as he would also say it of a thermodynamic ensemble, "This situation is highly organized, it is not characterized by a large degree of randomness or of choice — that is to say, the information (or the entropy) is low." We will return to this point later, for unless I am quite mistaken, it is an important aspect of the more general significance of this theory.

Having calculated the entropy (or the information, or the freedom of choice) of a certain information source, one can compare this to the maximum value this entropy could have, subject only to the condition that the source continue to employ the same symbols. The ratio of the actual to the maximum entropy is called the *relative entropy* of the source. If the relative entropy of a certain source is, say .8, this roughly means that this source is, in its choice of symbols to form a message, about 80 per cent as free as it could possibly be with these same symbols. One minus the relative entropy is called the *redundancy*. This is the fraction of the structure of the message which is determined not by the free choice of the sender, but rather by the accepted statistical rules governing the use of the symbols in question. It is sensibly called redundancy, for this fraction of the message is in fact redundant in something close to the ordinary sense; that is to say, this fraction of the message is unnecessary (and hence repetitive or redundant) in the sense that if it were missing the message would still be essentially complete, or at least could be completed.

It is most interesting to note that the redundancy of English is just about 50 per cent,[4] so that about half of the letters or words we choose in writing or speaking are under our free choice, and about half (although we are not ordinarily aware of it) are really controlled by the statistical structure of the language.

[4] The 50 per cent estimate accounts only for statistical structure out to about eight letters, so that the ultimate value is presumably a little higher.

Apart from more serious implications, which again we will postpone to our final discussion, it is interesting to note that a language must have at least 50 per cent of real freedom (or relative entropy) in the choice of letters if one is to be able to construct satisfactory crossword puzzles. If it has complete freedom, then every array of letters is a crossword puzzle. If it has only 20 per cent of freedom, then it would be impossible to construct crossword puzzles in such complexity and number as would make the game popular. Shannon has estimated that if the English language had only about 30 per cent redundancy, then it would be possible to construct three-dimensional crossword puzzles.

Before closing this section on information, it should be noted that the real reason that Level A analysis deals with a concept of information which characterizes the whole statistical nature of the information source, and is not concerned with the individual messages (and not at all directly concerned with the meaning of the individual messages) is that from the point of view of engineering, a communication system must face the problem of handling any message that the source can produce. If it is not possible or practicable to design a system which can handle everything perfectly, then the system should be designed to handle well the jobs it is most likely to be asked to do, and should resign itself to be less efficient for the rare task. This sort of consideration leads at once to the necessity of characterizing the statistical nature of the whole ensemble of messages which a given kind of source can and will produce. And *information*, as used in communication theory, does just this.

Although it is not at all the purpose of this paper to be concerned with mathematical details, it nevertheless seems essential to have as good an understanding as possible of the entropy-like expression which measures information. If one is concerned, as in a simple case, with a set of n independent symbols, or a set of n independent complete messages for that matter, whose probabilities of choice are $p_1, p_2 \cdots p_n$, then the actual expression for the information is

$$H = - [p_1 \log p_1 + p_2 \log p_2 + \cdots + p_n \log p_n],$$

or

$$H = - \Sigma \, p_i \log p_i.$$

Where[5] the symbol Σ indicates, as is usual in mathematics, that one is to sum all terms like the typical one, $p_i \log p_i$, written as a defining sample.

This looks a little complicated; but let us see how this expression behaves in some simple cases.

Suppose first that we are choosing only between two possible messages, whose probabilities are then p_1 for the first and $p_2 = 1 - p_1$ for the other. If one reckons, for this case, the numerical value of H, it turns out that H has its largest value, namely one, when the two messages are equally probable; that is to say when $p_1 = p_2 = \frac{1}{2}$; that is to say, when one is completely free to choose between the two messages. Just as soon as one message becomes more probable than the other (p_1 greater than p_2, say), the value of H decreases. And when one message is very probable (p_1 almost one and p_2 almost zero, say), the value of H is very small (almost zero).

In the limiting case where one probability is unity (certainty) and all the others zero (impossibility), then H is zero (no uncertainty at all — no freedom of choice — no information).

Thus H is largest when the two probabilities are equal (i.e., when one is completely free and unbiased in the choice), and reduces to zero when one's freedom of choice is gone.

The situation just described is in fact typical. If there are many, rather than two, choices, then H is largest when the probabilities of the various choices are as nearly equal as circumstances permit — when one has as much freedom as possible in making a choice, being as little as possible driven toward some certain choices which have more than their share of probability. Suppose, on the other hand, that one choice has a probability near one so that all the other choices have probabilities near zero. This is clearly a situation in which one is heavily influenced toward one particular choice, and hence has little freedom of choice. And H in such a case does calculate to have a very small value — the information (the freedom of choice, the uncertainty) is low.

When the number of cases is fixed, we have just seen that then

---

[5] Do not worry about the minus sign. Any probability is a number less than or equal to one, and the logarithms of numbers less than one are themselves negative. Thus the minus sign is necessary in order that H be in fact positive.

the information is the greater, the more nearly equal are the probabilities of the various cases. There is another important way of increasing H, namely by increasing the number of cases. More accurately, if all choices are equally likely, the more choices there are, the larger H will be. There is more "information" if you select freely out of a set of fifty standard messages, than if you select freely out of a set of twenty-five.

2.3. Capacity of a Communication Channel

After the discussion of the preceding section, one is not surprised that the capacity of a channel is to be described not in terms of the number of *symbols* it can transmit, but rather in terms of the *information* it transmits. Or better, since this last phrase lends itself particularly well to a misinterpretation of the word information, the capacity of a channel is to be described in terms of its ability to transmit what is produced out of source of a given information.

If the source is of a simple sort in which all symbols are of the same time duration (which is the case, for example, with teletype), if the source is such that each symbol chosen represents s bits of information (being freely chosen from among 2^s symbols), and if the channel can transmit, say n symbols per second, then the capacity of C of the channel is defined to be ns bits per second.

In a more general case, one has to take account of the varying lengths of the various symbols. Thus the general expression for capacity of a channel involves the logarithm of the numbers of symbols of certain time duration (which introduces, of course, the idea of *information* and corresponds to the factor s in the simple case of the preceding paragraph); and also involves the number of such symbols handled (which corresponds to the factor n of the preceding paragraph). Thus in the general case, capacity measures not the number of symbols transmitted per second, but rather the amount of information transmitted per second, using bits per second as its unit.

2.4. Coding

At the outset it was pointed out that the *transmitter* accepts the *message* and turns it into something called the *signal,* the latter being what actually passes over the channel to the *receiver.*

The transmitter, in such a case as telephony, merely changes the audible voice signal over into something (the varying electrical current on the telephone wire) which is at once clearly different but clearly equivalent. But the transmitter may carry out a much more complex operation on the message to produce the signal. It could, for example, take a written message and use some code to encipher this message into, say a sequence of numbers; these numbers then being sent over the channel as the signal.

Thus one says, in general, that the function of the transmitter is to *encode*, and that of the receiver to *decode*, the message. The theory provides for very sophisticated transmitters and receivers — such, for example, as possess "memories," so that the way they encode a certain symbol of the message depends not only upon this one symbol, but also upon previous symbols of the message and the way they have been encoded.

We are now in a position to state the fundamental theorem, produced in this theory, for a noiseless channel transmitting discrete symbols. This theorem relates to a communication channel which has a capacity of C bits per second, accepting signals from a source of entropy (or information) of H bits per second. The theorem states that by devising proper coding procedures for the transmitter it is possible to transmit symbols over the channel at an average rate[6] which is nearly C/H, but which, no matter how clever the coding, can never be made to exceed C/H.

The significance of this theorem is to be discussed more usefully a little later, when we have the more general case when noise is present. For the moment, though, it is important to notice the critical role which coding plays.

Remember that the entropy (or information) associated with the process which generates messages or signals is determined by the statistical character of the process — by the various probabilities for arriving at message situations and for choosing, when in those situations the next symbols. The statistical nature of *messages* is entirely determined by the character of the source.

[6] We remember that the capacity C involves the idea of information transmitted per second, and is thus measured in bits per second. The entropy H here measures information per symbol, so that the ratio of C to H measures symbols per second.

But the statistical character of the *signal* as actually transmitted by a channel, and hence the entropy in the channel, is determined both by what one attempts to feed into the channel and by the capabilities of the channel to handle different signal situations. For example, in telegraphy, there have to be spaces between dots and dots, between dots and dashes, and between dashes and dashes, or the dots and dashes would not be recognizable.

Now it turns out that when a channel does have certain constraints of this sort, which limit complete signal freedom, there are certain statistical signal characteristics which lead to a signal entropy which is larger than it would be for any other statistical signal structure, and in this important case, the signal entropy is exactly equal to the channel capacity.

In terms of these ideas, it is now possible precisely to characterize the most efficient kind of coding. The best transmitter, in fact, is that which codes the message in such a way that the signal has just those optimum statistical characteristics which are best suited to the channel to be used — which in fact maximize the signal (or one may say, the channel) entropy and make it equal to the capacity C of the channel.

This kind of coding leads, by the fundamental theorem above, to the maximum rate C/H for the transmission of symbols. But for this gain in transmission rate, one pays a price. For rather perversely it happens that as one makes the coding more and more nearly ideal, one is forced to longer and longer delays *in the process of coding*. Part of this dilemma is met by the fact that in electronic equipment "long" may mean an exceedingly small fraction of a second, and part by the fact that one makes a compromise, balancing the gain in transmission rate against loss of coding time.

2.5. Noise

How does noise affect information? Information is, we must steadily remember, a measure of one's freedom of choice in selecting a message. The greater this freedom of choice, and hence the greater the information, the greater is the uncertainty that the message actually selected is some particular one. Thus greater freedom of choice, greater uncertainty, greater information go hand in hand.

If noise is introduced, then the received message contains certain distortions, certain errors, certain extraneous material, that would certainly lead one to say that the received message exhibits, because of the effects of the noise, an increased uncertainty. But if the uncertainty is increased, the information is increased, and this sounds as though the noise were beneficial!

It is generally true that when there is noise, the received signal exhibits greater information — or better, the received signal is selected out of a more varied set than is the transmitted signal. This is a situation which beautifully illustrates the semantic trap into which one can fall if he does not remember that "information" is used here with a special meaning that measures freedom of choice and hence uncertainty as to what choice has been made. It is therefore possible for the word information to have either good or bad connotations. Uncertainty which arises by virtue of freedom of choice on the part of the sender is desirable uncertainty. Uncertainty which arises because of errors or because of the influence of noise is undesirable uncertainty.

It is thus clear where the joker is in saying that the received signal has more information. Some of this information is spurious and undesirable and has been introduced via the noise. To get the useful information in the received signal we must subtract out this spurious portion.

Before we can clear up this point we have to stop for a little detour. Suppose one has two sets of symbols, such as the message symbols generated by the information source, and the signal symbols which are actually received. The probabilities of these two sets of symbols are interrelated, for clearly the probability of receiving a certain symbol depends upon what symbol was sent. With no errors from noise or from other causes, the received signals would correspond precisely to the message symbols sent; and in the presence of possible error, the probabilities for received symbols would obviously be loaded heavily on those which correspond, or closely correspond, to the message symbols sent.

Now in such a situation one can calculate what is called the entropy of one set of symbols relative to the other. Let us, for example, consider the entropy of the message relative to the signal. It is unfortunate that we cannot understand the issues involved here without going into some detail. Suppose for the moment that one knows that a certain signal symbol has actually been received. Then each *message* symbol takes on a certain probability — relatively large for the symbol identical with or the symbols similar to the one received, and relatively small for all others. Using this set of probabilities, one calculates a tentative entropy value. This is the message entropy on the assumption of a definite known received or signal symbol. Under any good conditions its value is low, since the probabilities involved are not spread around rather evenly on the various cases, but are heavily loaded on one or a few cases. Its value would be zero (see page 13) in any case where noise was completely absent, for then, the signal symbol being known, all message probabilities would be zero except for one symbol (namely the one received), which would have a probability of unity.

For each assumption as to the signal symbol received, one can calculate one of these tentative message entropies. Calculate all of them, and then average them, weighting each one in accordance with the probability of the signal symbol assumed in calculating it. Entropies calculated in this way, when there are two sets of symbols to consider, are called *relative entropies*. The particular one just described is the entropy of the message relative to the signal, and Shannon has named this also the *equivocation*.

From the way this equivocation is calculated, we can see what its significance is. It measures the *average uncertainty in the message when the signal is known*. If there were no noise, then there would be no uncertainty concerning the message if the signal is known. If the information source has any residual uncertainty after the signal is known, then this must be undesirable uncertainty due to noise.

The discussion of the last few paragraphs centers around the quantity "the average uncertainty in the message source when the received signal is known." It can equally well be phrased in terms of the similar quantity "the average uncertainty concerning the received signal when the message sent is known." This latter uncertainty would, of course, also be zero if there were no noise.

As to the interrelationship of these quantities, it is easy to prove that

$$H(x) - H_y(x) = H(y) - H_x(y)$$

where $H(x)$ is the entropy or information of the source of messages; $H(y)$ the entropy or information of received signals; $H_y(x)$ the equivocation, or the uncertainty in the message source if the signal be known; $H_x(y)$ the uncertainty in the received signals if the messages sent be known, or the spurious part of the received signal information which is due to noise. The right side of this equation is the useful information which is transmitted in spite of the bad effect of the noise.

It is now possible to explain what one means by the capacity C of a noisy channel. It is, in fact, defined to be equal to the maximum rate (in bits per second) at which useful information (i.e., total uncertainty minus noise uncertainty) can be transmitted over the channel.

Why does one speak, here, of a "maximum" rate? What can one do, that is, to make this rate larger or smaller? The answer is that one can affect this rate by choosing a source whose statistical characteristics are suitably related to the restraints imposed by the nature of the channel. That is, one can maximize the rate of transmitting useful information by using proper coding (see pages 16-17).

And now, finally, let us consider the fundamental theorem for a noisy channel. Suppose that this noisy channel has, in the sense just described, a capacity C, suppose it is accepting from an information source characterized by an entropy of $H(x)$ bits per second, the entropy of the received signals being $H(y)$ bits per second. If the channel capacity C is equal to or larger than $H(x)$, then by devising appropriate coding systems, the output of the source can be transmitted over the channel with as little error as one pleases. However small a frequency of error you specify, there is a code which meets the demand. But if the channel capacity C is less than $H(x)$, the entropy of the source from which it accepts messages, then it is impossible to devise codes which reduce the error frequency as low as one may please.

However clever one is with the coding process, it will always be true that after the signal is received there remains some undesirable (noise) uncertainty about what the message was; and this undesirable uncertainty — this equivocation — will always be equal to or greater than $H(x) - C$. Furthermore, there is always at least one code which is capable of reducing this undesirable uncertainty, concerning the message, down to a value which exceeds $H(x) - C$ by an arbitrarily small amount.

The most important aspect, of course, is that the minimum undesirable or spurious uncertainties cannot be reduced further, no matter how complicated or appropriate the coding process. This powerful theorem gives a precise and almost startlingly simple description of the utmost dependability one can ever obtain from a communication channel which operates in the presence of noise.

One practical consequence, pointed out by Shannon, should be noted. Since English is about 50 per cent redundant, it would be possible to save about one-half the time of ordinary telegraphy by a proper encoding process, *provided* one were going to transmit over a noiseless channel. When there is noise on a channel, however, there is some real advantage in not using a coding process that eliminates all of the redundancy. For the remaining redundancy helps combat the noise. This is very easy to see, for just because of the fact that the redundancy of English is high, one has, for example, little or no hesitation about correcting errors in spelling that have arisen during transmission.

2.6. Continuous Messages

Up to this point we have been concerned with messages formed out of discrete symbols, as words are formed of letters, sentences of words, a melody of notes, or a halftone picture of a finite number of discrete spots. What happens to the theory if one considers continuous messages, such as the speaking voice with its continuous variation of pitch and energy?

Very roughly one may say that the extended theory is somewhat more difficult and complicated mathematically, but not essentially different. Many of the above statements for the discrete case require no modification, and others require only minor change.

One circumstance which helps a good deal is the following. As a practical matter, one is always interested in a continuous signal which is built up of simple harmonic constituents *of not all frequencies,* but rather of frequencies which lie wholly within a band from zero frequency to, say, a frequency of W cycles per second. Thus although the human voice does contain higher frequencies, very satisfactory communication can be achieved over a telephone channel that handles frequencies only up to, say four thousand. With frequencies up to ten or twelve thousand, high fidelity radio transmission of symphonic music is possible, etc.

There is a very convenient mathematical theorem which states that a continuous signal, T seconds in duration and band-limited in frequency to the range from 0 to W, can be *completely specified* by stating $2TW$ numbers. This is really a remarkable theorem. Ordinarily a continuous curve can be only approximately characterized by stating any finite number of points through which it passes, and an infinite number would in general be required for complete information about the curve. But if the curve is built up out of simple harmonic constituents of a limited number of frequencies, as a complex sound is built up out of a limited number of pure tones, then a finite number of parameters is all that is necessary. This has the powerful advantage of reducing the character of the communication problem for continuous signals from a complicated situation where one would have to deal with an infinite number of variables to a considerably simpler situation where one deals with a finite (though large) number of variables.

In the theory for the continuous case there are developed formulas which describe the maximum capacity C of a channel of frequency bandwidth W, when the average *power* used in transmitting is P, the channel being subject to a noise of power N, this noise being "white thermal noise" of a special kind which Shannon defines. This white thermal noise is itself band limited in frequency, and the amplitudes of the various frequency constituents are subject to a normal (Gaussian) probability distribution. Under these circumstances, Shannon obtains the theorem, again really quite remarkable in its simplicity and its scope, that it is possible, by the best coding, to transmit binary digits at the rate of

$$W \log_2 \frac{P + N}{N}$$

bits per second and have an arbitrarily low frequency of error. But this rate cannot possibly be exceeded, no matter how clever the coding, without giving rise to a definite frequency of errors. For the case of arbitrary noise, rather than the special "white thermal" noise assumed above, Shannon does not succeed in deriving one explicit formula for channel capacity, but does obtain useful upper and lower limits for channel capacity. And he also derives limits for channel capacity when one specifies not the average power of the transmitter, but rather the peak instantaneous power.

Finally it should be stated that Shannon obtains results which are necessarily somewhat less specific, but which are of obviously deep and sweeping significance, which, for a general sort of continuous message or signal, characterize the fidelity of the received message, and the concepts of rate at which a source generates information, rate of transmission, and channel capacity, all of these being relative to certain fidelity requirements.

3
The Interrelationship of the Three Levels of Communication Problems

3.1. Introductory

In the first section of this paper it was suggested that there are three levels at which one may consider the general communication problem. Namely, one may ask:

LEVEL A. How accurately can the symbols of communication be transmitted?

LEVEL B. How precisely do the transmitted symbols convey the desired meaning?

LEVEL C. How effectively does the received meaning affect conduct in the desired way?

It was suggested that the mathematical theory of communication, as developed by Shannon, Wiener, and others, and particularly the more definitely engineering theory treated by Shannon, although ostensibly applicable only to Level A problems, actually is helpful and suggestive for the level B and C problems.

We then took a look, in section 2, at what this mathematical theory is, what concepts it develops, what results it has obtained. It is the purpose of this concluding section to review the situation, and see to what extent and in what terms the original section was justified in indicating that the progress made at Level A is capable of contributing to levels B and C, was justified in indicating that the interrelation of the three levels is so considerable that one's final conclusion may be that the separation into the three levels is really artificial and undesirable.

3.2. Generality of the Theory at Level A

The obvious first remark, and indeed the remark that carries the major burden of the argument, is that the mathematical theory is exceedingly general in its scope, fundamental in the problems it treats, and of classic simplicity and power in the results it reaches.

This is a theory so general that one does not need to say what kinds of symbols are being considered — whether written letters or words, or musical notes, or spoken words, or symphonic music, or pictures. The theory is deep enough so that the relationships it reveals indiscriminately apply to all these and to other forms of communication. This means, of course, that the theory is sufficiently imaginatively motivated so that it is dealing with the real inner core of the communication problem — with those basic relationships which hold in general, no matter what special form the actual case may take.

It is an evidence of this generality that the theory contributes importantly to, and in fact is really the basic theory of cryptography which is, of course, a form of coding. In a similar way, the theory contributes to the problem of translation from one language to another, although the complete story here clearly requires consideration of meaning, as well as of information. Similarly, the ideas developed in this work connect so closely with the problem of the logical design of great computers that it is no surprise that Shannon has just written a paper on the design of a computer which would be capable of playing a skillful game of chess. And it is of further direct pertinence to the present contention that this paper closes with the remark that either one must say that such a computer "thinks," or one must substantially modify the conventional implication of the verb "to think."

As a second point, it seems clear that an important contribution has been made to any possible general theory of communication by the formalization on which the present theory is based. It seems at first obvious to diagram a communication system as it is done at the outset of this theory; but this breakdown of the situation must be very deeply sensible and appropriate, as one becomes convinced when he sees how smoothly and generally this viewpoint leads to central issues. It is almost certainly true that a consideration of communication on levels B and C will require additions to the schematic diagram on page 7, but it seems equally likely that what is required are minor additions, and no real revision.

Thus when one moves to levels B and C, it may prove to be essential to take account of the statistical characteristics of the destination. One can imagine, as an addition to the diagram, another box labeled "Semantic Receiver" interposed between the engineering receiver (which changes signals to messages) and the destination. This semantic receiver subjects the message to a second decoding, the demand on this one being that it must match the statistical *semantic* characteristics of the message to the statistical semantic capacities of the totality of receivers, or of that subset of receivers which constitute the audience one wishes to affect.

Similarly one can imagine another box in the diagram which, inserted between the information source and the transmitter, would be labeled "semantic noise," the box previously labeled as simply "noise" now being labeled "engineering noise." From this source is imposed into the signal the perturbations or distortions of meaning which are not intended by the source but which inescapably affect the destination. And the problem of semantic decoding must take this semantic noise into account. It is also possible to think of an adjustment of original message so that the sum of message meaning plus semantic noise is equal to the desired total message meaning at the destination.

Thirdly, it seems highly suggestive for the problem at all levels that error and confusion arise and fidelity decreases, when, no matter how good the coding, one tries to crowd too much over a channel (i.e., $H > C$). Here again a general theory at all levels will surely have to take into account not only the capacity of the channel but also (even the words are right!) the capacity of the audience. If one tries to overcrowd the capacity of the audience, it is probably true, by direct analogy, that you do not, so to speak, fill the audience up and then waste only the remainder by spilling. More likely, and again by direct analogy, if you overcrowd the capacity of the audience you force a general and inescapable error and confusion.

Fourthly, it is hard to believe that levels B and C do not have much to learn from, and do not have the approach to their problems usefully oriented by, the development in this theory of the entropic ideas in relation to the concept of information.

The concept of information developed in this theory at first seems disappointing and bizarre — disappointing because it has nothing to do with meaning, and bizarre because it deals not with a single message but rather with the statistical character of a whole ensemble of messages, bizarre also because in these statistical terms the two words *information* and *uncertainty* find themselves to be partners.

I think, however, that these should be only temporary reactions; and that one should say, at the end, that this analysis has so penetratingly cleared the air that one is now, perhaps for the first time, ready for a real theory of meaning. An engineering communication theory is just like a very proper and discreet girl accepting your telegram. She pays no attention to the meaning, whether it be sad, or joyous, or embarrassing. But she must be prepared to deal with all that come to her desk. This idea that a communication system ought to try to deal with all possible messages, and that the intelligent way to try is to base design on the statistical character of the source, is surely not without significance for communication in general. Language must be designed (or developed) with a view to the totality of things that man may wish to say; but not being able to accomplish everything, it too should do as well as possible as often as possible. That is to say, it too should deal with its task statistically.

The concept of the information to be associated with a source leads directly, as we have seen, to a study of the statistical structure of language; and this study reveals about the English language, as an example, information which seems surely significant to students of every phase of language and communication. The idea of utilizing the powerful body of theory concerning Markoff processes seems particularly promising for semantic studies, since this theory is specifically adapted to handle one of the most significant but difficult aspects of meaning, namely the influence of context. One has the vague feeling that information and meaning may prove to be something like a pair of canonically conjugate variables in quantum theory, they being subject to some joint restriction that condemns a person to the sacrifice of the one as he insists on having much of the other.

Or perhaps meaning may be shown to be analogous to one of the quantities on which the entropy of a thermodynamic ensemble depends. The appearance of entropy in the theory, as was remarked earlier, is surely most interesting and significant. Eddington has already been quoted in this connection, but there is another passage in "The Nature of the Physical World" which seems particularly apt and suggestive:

> Suppose that we were asked to arrange the following in two categories — *distance, mass, electric force, entropy, beauty, melody.*
> I think there are the strongest grounds for placing entropy alongside beauty and melody, and not with the first three. Entropy is only found when the parts are viewed in association, and it is by viewing or hearing the parts in association that beauty and melody are discerned. All three are features of arrangement. It is a pregnant thought that one of these three associates should be able to figure as a commonplace quantity of science. The reason why this stranger can pass itself off among the aborigines of the physical world is that it is able to speak their language, viz., the language of arithmetic.

I feel sure that Eddington would have been willing to include the word *meaning* along with beauty and melody; and I suspect he would have been thrilled to see, in this theory, that entropy not only speaks the language of arithmetic; it also speaks the language of language.

THE MATHEMATICAL THEORY OF COMMUNICATION

By Claude E. Shannon

Introduction

The recent development of various methods of modulation such as PCM and PPM which exchange bandwidth for signal-to-noise ratio has intensified the interest in a general theory of communication. A basis for such a theory is contained in the important papers of Nyquist[1] and Hartley[2] on this subject. In the present paper we will extend the theory to include a number of new factors, in particular the effect of noise in the channel, and the savings possible due to the statistical structure of the original message and due to the nature of the final destination of the information.

The fundamental problem of communication is that of reproducing at one point either exactly or approximately a message selected at another point. Frequently the messages have *meaning;* that is they refer to or are correlated according to some system with certain physical or conceptual entities. These semantic aspects of communication are irrelevant to the engineering problem. The significant aspect is that the actual message is one *selected from a set* of possible messages. The system must be designed to operate for each possible selection, not just the one which will actually be chosen since this is unknown at the time of design.

[1] Nyquist, H., "Certain Factors Affecting Telegraph Speed," *Bell System Technical Journal,* April 1924, p. 324; "Certain Topics in Telegraph Transmission Theory," *A.I.E.E. Trans.,* v. 47, April 1928, p. 617.
[2] Hartley, R. V. L., "Transmission of Information," *Bell System Technical Journal,* July 1928, p. 535.

If the number of messages in the set is finite then this number or any monotonic function of this number can be regarded as a measure of the information produced when one message is chosen from the set, all choices being equally likely. As was pointed out by Hartley the most natural choice is the logarithmic function. Although this definition must be generalized considerably when we consider the influence of the statistics of the message and when we have a continuous range of messages, we will in all cases use an essentially logarithmic measure.

The logarithmic measure is more convenient for various reasons:

1. It is practically more useful. Parameters of engineering importance such as time, bandwidth, number of relays, etc., tend to vary linearly with the logarithm of the number of possibilities. For example, adding one relay to a group doubles the number of possible states of the relays. It adds 1 to the base 2 logarithm of this number. Doubling the time roughly squares the number of possible messages, or doubles the logarithm, etc.

2. It is nearer to our intuitive feeling as to the proper measure. This is closely related to (1) since we intuitively measure entities by linear comparison with common standards. One feels, for example, that two punched cards should have twice the capacity of one for information storage, and two identical channels twice the capacity of one for transmitting information.

3. It is mathematically more suitable. Many of the limiting operations are simple in terms of the logarithm but would require clumsy restatement in terms of the number of possibilities.

The choice of a logarithmic base corresponds to the choice of a unit for measuring information. If the base 2 is used the resulting units may be called binary digits, or more briefly *bits*, a word suggested by J. W. Tukey. A device with two stable positions, such as a relay or a flip-flop circuit, can store one bit of information. N such devices can store N bits, since the total number of possible states is 2^N and $\log_2 2^N = N$. If the base 10 is used the units may be called decimal digits. Since

$$\log_2 M = \log_{10} M / \log_{10} 2$$
$$= 3.32 \log_{10} M,$$

a decimal digit is about $3\frac{1}{3}$ bits. A digit wheel on a desk computing machine has ten stable positions and therefore has a storage capacity of one decimal digit. In analytical work where integration and differentiation are involved the base e is sometimes useful. The resulting units of information will be called natural units. Change from the base a to base b merely requires multiplication by $\log_b a$.

By a communication system we will mean a system of the type indicated schematically in Fig. 1. It consists of essentially five parts:

1. An *information source* which produces a message or sequence of messages to be communicated to the receiving terminal. The message may be of various types: (a) A sequence of letters as in a telegraph or teletype system; (b) A single function of time $f(t)$ as in radio or telephony; (c) A function of time and other variables as in black and white television — here the message may be thought of as a function $f(x, y, t)$ of two space coordinates and time, the light intensity at point (x, y) and time t on a pickup tube plate; (d) Two or more functions of time, say $f(t), g(t), h(t)$ — this is the case in "three-dimensional" sound transmission or if the system is intended to service several individual channels in multiplex; (e) Several functions of several variables — in color television the message consists of three functions $f(x, y, t), g(x, y, t), h(x, y, t)$ defined in a three-dimensional continuum — we may also think of these three functions as components of a vector field defined in the region — similarly, several black and white television sources would produce "messages" consisting of a number of functions of three variables; (f) Various combinations also occur, for example in television with an associated audio channel.

2. A *transmitter* which operates on the message in some way to produce a signal suitable for transmission over the channel. In telephony this operation consists merely of changing sound pressure into a proportional electrical current. In telegraphy we have an encoding operation which produces a sequence of dots, dashes and spaces on the channel corresponding to the message. In a multiplex PCM system the different speech functions must be sampled, compressed, quantized and encoded, and finally inter-

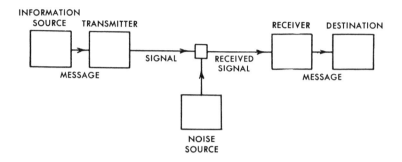

Fig. 1. — Schematic diagram of a general communication system.

leaved properly to construct the signal. Vocoder systems, television and frequency modulation are other examples of complex operations applied to the message to obtain the signal.

3. The *channel* is merely the medium used to transmit the signal from transmitter to receiver. It may be a pair of wires, a coaxial cable, a band of radio frequencies, a beam of light, etc. During transmission, or at one of the terminals, the signal may be perturbed by noise. This is indicated schematically in Fig. 1 by the noise source acting on the transmitted signal to produce the received signal.

4. The *receiver* ordinarily performs the inverse operation of that done by the transmitter, reconstructing the message from the signal.

5. The *destination* is the person (or thing) for whom the message is intended.

We wish to consider certain general problems involving communication systems. To do this it is first necessary to represent the various elements involved as mathematical entities, suitably idealized from their physical counterparts. We may roughly classify communication systems into three main categories: discrete, continuous and mixed. By a discrete system we will mean one in which both the message and the signal are a sequence of discrete symbols. A typical case is telegraphy where the message is a sequence of letters and the signal a sequence of dots, dashes and spaces. A continuous system is one in which the message and signal are both treated as continuous functions, e.g., radio or television. A mixed system is one in which both discrete and continuous variables appear, e.g., PCM transmission of speech.

We first consider the discrete case. This case has applications not only in communication theory, but also in the theory of computing machines, the design of telephone exchanges and other fields. In addition the discrete case forms a foundation for the continuous and mixed cases which will be treated in the second half of the paper.

I

Discrete Noiseless Systems

1. The Discrete Noiseless Channel

Teletype and telegraphy are two simple examples of a discrete channel for transmitting information. Generally, a discrete channel will mean a system whereby a sequence of choices from a finite set of elementary symbols $S_1 \cdots S_n$ can be transmitted from one point to another. Each of the symbols S_i is assumed to have a certain duration in time t_i seconds (not necessarily the same for different S_i, for example the dots and dashes in telegraphy). It is not required that all possible sequences of the S_i be capable of transmission on the system; certain sequences only may be allowed. These will be possible signals for the channel. Thus in telegraphy suppose the symbols are: (1) A dot, consisting of line closure for a unit of time and then line open for a unit of time; (2) A dash, consisting of three time units of closure and one unit open; (3) A letter space consisting of, say, three units of line open; (4) A word space of six units of line open. We might place the restriction on allowable sequences that no spaces follow each other (for if two letter spaces are adjacent, they are identical with a word space). The question we now consider is how one can measure the capacity of such a channel to transmit information.

In the teletype case where all symbols are of the same duration, and any sequence of the 32 symbols is allowed, the answer is easy. Each symbol represents five bits of information. If the system transmits n symbols per second it is natural to say that the channel has a capacity of $5n$ bits per second. This does not mean that the teletype channel will always be transmitting information at this rate — this is the maximum possible rate and whether or not the actual rate reaches this maximum depends on the source of information which feeds the channel, as will appear later.

In the more general case with different lengths of symbols and constraints on the allowed sequences, we make the following definition: The capacity C of a discrete channel is given by

$$C = \lim_{T \to \infty} \frac{\log N(T)}{T}$$

where $N(T)$ is the number of allowed signals of duration T.

It is easily seen that in the teletype case this reduces to the previous result. It can be shown that the limit in question will exist as a finite number in most cases of interest. Suppose all sequences of the symbols $S_1, \cdots, S_n$ are allowed and these symbols have durations $t_1, \cdots, t_n$. What is the channel capacity? If $N(t)$ represents the number of sequences of duration t we have

$$N(t) = N(t - t_1) + N(t - t_2) + \cdots + N(t - t_n).$$

The total number is equal to the sum of the numbers of sequences ending in $S_1, S_2, \cdots, S_n$ and these are $N(t - t_1)$, $N(t - t_2)$, $\cdots$, $N(t - t_n)$, respectively. According to a well-known result in finite differences, $N(t)$ is then asymptotic for large t to AX_0^t where A is constant and X_0 is the largest real solution of the characteristic equation:

$$X^{-t_1} + X^{-t_2} + \cdots + X^{-t_n} = 1$$

and therefore

$$C = \lim_{T \to \infty} \frac{\log AX_0^T}{T} = \log X_0.$$

In case there are restrictions on allowed sequences we may still often obtain a difference equation of this type and find C from the characteristic equation. In the telegraphy case mentioned above

$$N(t) = N(t - 2) + N(t - 4) + N(t - 5) + N(t - 7)$$
$$+ N(t - 8) + N(t - 10)$$

as we see by counting sequences of symbols according to the last or next to the last symbol occurring. Hence C is $-\log \mu_0$ where μ_0 is the positive root of $1 = \mu^2 + \mu^4 + \mu^5 + \mu^7 + \mu^8 + \mu^{10}$. Solving this we find $C = 0.539$.

A very general type of restriction which may be placed on allowed sequences is the following: We imagine a number of possible states $a_1, a_2, \cdots, a_m$. For each state only certain symbols from the set $S_1, \cdots, S_n$ can be transmitted (different subsets for the different states). When one of these has been transmitted the state changes to a new state depending both on the old state and the particular symbol transmitted. The telegraph case is a simple example of this. There are two states depending on whether or not a space was the last symbol transmitted. If so, then only a dot or a dash can be sent next and the state always changes. If not, any symbol can be transmitted and the state changes if a space is sent, otherwise it remains the same. The conditions can be indicated in a linear graph as shown in Fig. 2. The junction points correspond to the states and the lines indicate the symbols possible in a state and the resulting state. In Appendix 1 it is shown that if the conditions on allowed sequences can be described in this form C will exist and can be calculated in accordance with the following result:

Theorem 1: Let $b_{ij}^{(s)}$ be the duration of the sth symbol which is allowable in state i and leads to stage j. Then the channel capacity C is equal to log W *where W is the largest real root of the determinantal equation:*

$$\left| \sum_s W^{-b_{ij}^{(s)}} - \delta_{ij} \right| = 0$$

where $\delta_{ij} = 1$ if $i = j$ and is zero otherwise.

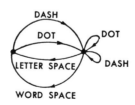

Fig. 2. — Graphical representation of the constraints on telegraph symbols.

For example, in the telegraph case (Fig. 2) the determinant is:

$$\begin{vmatrix} -1 & (W^{-2} + W^{-4}) \\ (W^{-3} + W^{-6}) & (W^{-2} + W^{-4} - 1) \end{vmatrix} = 0.$$

On expansion this leads to the equation given above for this set of constraints.

2. The Discrete Source of Information

We have seen that under very general conditions the logarithm of the number of possible signals in a discrete channel increases linearly with time. The capacity to transmit information can be specified by giving this rate of increase, the number of bits per second required to specify the particular signal used.

We now consider the information source. How is an information source to be described mathematically, and how much information in bits per second is produced in a given source? The main point at issue is the effect of statistical knowledge about the source in reducing the required capacity of the channel, by the use of proper encoding of the information. In telegraphy, for example, the messages to be transmitted consist of sequences of letters. These sequences, however, are not completely random. In general, they form sentences and have the statistical structure of, say, English. The letter E occurs more frequently than Q, the sequence TH more frequently than XP, etc. The existence of this structure allows one to make a saving in time (or channel capacity) by properly encoding the message sequences into signal sequences. This is already done to a limited extent in telegraphy by using the shortest channel symbol, a dot, for the most common English letter E; while the infrequent letters, Q, X, Z are represented by longer sequences of dots and dashes. This idea is carried still further in certain commercial codes where common words and phrases are represented by four- or five-letter code groups with a considerable saving in average time. The standardized greeting and anniversary telegrams now in use extend this to the point of encoding a sentence or two into a relatively short sequence of numbers.

We can think of a discrete source as generating the message, symbol by symbol. It will choose successive symbols according to certain probabilities depending, in general, on preceding choices as well as the particular symbols in question. A physical system, or a mathematical model of a system which produces such a sequence of symbols governed by a set of probabilities, is known as a stochastic process.[3] We may consider a discrete source, therefore, to be represented by a stochastic process. Conversely, any stochastic process which produces a discrete sequence of symbols chosen from a finite set may be considered a discrete source. This will include such cases as:

1. Natural written languages such as English, German, Chinese.

2. Continuous information sources that have been rendered discrete by some quantizing process. For example, the quantized speech from a PCM transmitter, or a quantized television signal.

3. Mathematical cases where we merely define abstractly a stochastic process which generates a sequence of symbols. The following are examples of this last type of source.

(A) Suppose we have five letters A, B, C, D, E which are chosen each with probability .2, successive choices being independent. This would lead to a sequence of which the following is a typical example.

B D C B C E C C C A D C B D D A A E C E E A A B B D A E E C A C E E B A E E C B C E A D.

This was constructed with the use of a table of random numbers.[4]

(B) Using the same five letters let the probabilities be .4, .1, .2, .2, .1, respectively, with successive choices independent. A typical message from this source is then:

A A A C D C B D C E A A D A D A C E D A E A D C A B E D A D D C E C A A A A A D.

(C) A more complicated structure is obtained if successive symbols are not chosen independently but their probabilities

---

[3] See, for example, S. Chandrasekhar, "Stochastic Problems in Physics and Astronomy," *Reviews of Modern Physics,* v. 15, No. 1, January 1943, p. 1.
[4] Kendall and Smith, *Tables of Random Sampling Numbers,* Cambridge, 1939.

depend on preceding letters. In the simplest case of this type
a choice depends only on the preceding letter and not on
ones before that. The statistical structure can then be described by a set of transition probabilities $p_i(j)$, the probability that letter i is followed by letter j. The indices i and
j range over all the possible symbols. A second equivalent
way of specifying the structure is to give the "digram"
probabilities $p(i,j)$, i.e., the relative frequency of the digram i j. The letter frequencies $p(i)$, (the probability of
letter i), the transition probabilities $p_i(j)$ and the digram
probabilities $p(i,j)$ are related by the following formulas:

$$p(i) = \sum_j p(i,j) = \sum_j p(j,i) = \sum_j p(j)p_j(i)$$

$$p(i,j) = p(i)p_i(j)$$

$$\sum_j p_i(j) = \sum_i p(i) = \sum_{i,j} p(i,j) = 1.$$

As a specific example suppose there are three letters A, B,
C with the probability tables:

$p_i(j)$		j			i	$p(i)$
		A	B	C		
	A	0	$\frac{4}{5}$	$\frac{1}{5}$	A	$\frac{9}{27}$
i	B	$\frac{1}{2}$	$\frac{1}{2}$	0	B	$\frac{16}{27}$
	C	$\frac{1}{2}$	$\frac{2}{5}$	$\frac{1}{10}$	C	$\frac{2}{27}$

$p(i,j)$		j		
		A	B	C
	A	0	$\frac{4}{15}$	$\frac{1}{15}$
i	B	$\frac{8}{27}$	$\frac{8}{27}$	0
	C	$\frac{1}{27}$	$\frac{4}{135}$	$\frac{1}{135}$

A typical message from this source is the following:

A B B A B A B A B A B A B A B A B B B A B B B B B A B A
B A B A B A B B B A C A C A B B A B B B B A B B A B
A C B B B A B A.

The next increase in complexity would involve trigram
frequencies but no more. The choice of a letter would depend on the preceding two letters but not on the message
before that point. A set of trigram frequencies $p(i, j, k)$ or
equivalently a set of transition probabilities $p_{ij}(k)$ would
be required. Continuing in this way one obtains successively
more complicated stochastic processes. In the general n-gram

case a set of n-gram probabilities $p(i_1, i_2, \cdots, i_n)$ or of transition probabilities $p_{i_1, i_2, \cdots, i_{n-1}}(i_n)$ is required to specify the statistical structure.

(D) Stochastic processes can also be defined which produce a text consisting of a sequence of "words." Suppose there are five letters A, B, C, D, E and 16 "words" in the language with associated probabilities:

.10 A	.16 BEBE	.11 CABED	.04 DEB
.04 ADEB	.04 BED	.05 CEED	.15 DEED
.05 ADEE	.02 BEED	.08 DAB	.01 EAB
.01 BADD	.05 CA	.04 DAD	.05 EE

Suppose successive "words" are chosen independently and are separated by a space. A typical message might be:

DAB EE A BEBE DEED DEB ADEE ADEE EE DEB BEBE BEBE BEBE ADEE BED DEED DEED CEED ADEE A DEED DEED BEBE CABED BEBE BED DAB DEED ADEB.

If all the words are of finite length this process is equivalent to one of the preceding type, but the description may be simpler in terms of the word structure and probabilities. We may also generalize here and introduce transition probabilities between words, etc.

These artificial languages are useful in constructing simple problems and examples to illustrate various possibilities. We can also approximate to a natural language by means of a series of simple artificial languages. The zero-order approximation is obtained by choosing all letters with the same probability and independently. The first-order approximation is obtained by choosing successive letters independently but each letter having the same probability that it has in the natural language.[5] Thus, in the first-order approximation to English, E is chosen with probability .12 (its frequency in normal English) and W with probability .02, but there is no influence between adjacent letters and no tendency to form the preferred digrams such as TH, ED, etc.

[5] Letter, digram and trigram frequencies are given in *Secret and Urgent* by Fletcher Pratt, Blue Ribbon Books, 1939. Word frequencies are tabulated in *Relative Frequency of English Speech Sounds*, G. Dewey, Harvard University Press, 1923.

In the second-order approximation, digram structure is introduced. After a letter is chosen, the next one is chosen in accordance with the frequencies with which the various letters follow the first one. This requires a table of digram frequencies $p_i(j)$. In the third-order approximation, trigram structure is introduced. Each letter is chosen with probabilities which depend on the preceding two letters.

3. The Series of Approximations to English

To give a visual idea of how this series of processes approaches a language, typical sequences in the approximations to English have been constructed and are given below. In all cases we have assumed a 27-symbol "alphabet," the 26 letters and a space.

1. Zero-order approximation (symbols independent and equiprobable).

   XFOML RXKHRJFFJUJ ZLPWCFWKCYJ FFJEYV-
   KCQSGHYD QPAAMKBZAACIBZLHJQD.

2. First-order approximation (symbols independent but with frequencies of English text).

   OCRO HLI RGWR NMIELWIS EU LL NBNESEBYA
   TH EEI ALHENHTTPA OOBTTVA NAH BRL.

3. Second-order approximation (digram structure as in English).

   ON IE ANTSOUTINYS ARE T INCTORE ST BE S
   DEAMY ACHIN D ILONASIVE TUCOOWE AT TEA-
   SONARE FUSO TIZIN ANDY TOBE SEACE CTISBE.

4. Third-order approximation (trigram structure as in English).

   IN NO IST LAT WHEY CRATICT FROURE BIRS
   GROCID PONDENOME OF DEMONSTURES OF
   THE REPTAGIN IS REGOACTIONA OF CRE.

5. First-order word approximation. Rather than continue with tetragram, $\cdots$, n-gram structure it is easier and better to jump at this point to word units. Here words are chosen independently but with their appropriate frequencies.

   REPRESENTING AND SPEEDILY IS AN GOOD APT
   OR COME CAN DIFFERENT NATURAL HERE HE

THE A IN CAME THE TO OF TO EXPERT GRAY COME TO FURNISHES THE LINE MESSAGE HAD BE THESE.

6. Second-order word approximation. The word transition probabilities are correct but no further structure is included.

THE HEAD AND IN FRONTAL ATTACK ON AN ENGLISH WRITER THAT THE CHARACTER OF THIS POINT IS THEREFORE ANOTHER METHOD FOR THE LETTERS THAT THE TIME OF WHO EVER TOLD THE PROBLEM FOR AN UNEXPECTED.

The resemblance to ordinary English text increases quite noticeably at each of the above steps. Note that these samples have reasonably good structure out to about twice the range that is taken into account in their construction. Thus in (3) the statistical process insures reasonable text for two-letter sequences, but four-letter sequences from the sample can usually be fitted into good sentences. In (6) sequences of four or more words can easily be placed in sentences without unusual or strained constructions. The particular sequence of ten words "attack on an English writer that the character of this" is not at all unreasonable. It appears then that a sufficiently complex stochastic process will give a satisfactory representation of a discrete source.

The first two samples were constructed by the use of a book of random numbers in conjunction with (for example 2) a table of letter frequencies. This method might have been continued for (3), (4) and (5), since digram, trigram and word frequency tables are available, but a simpler equivalent method was used. To construct (3) for example, one opens a book at random and selects a letter at random on the page. This letter is recorded. The book is then opened to another page and one reads until this letter is encountered. The succeeding letter is then recorded. Turning to another page this second letter is searched for and the succeeding letter recorded, etc. A similar process was use for (4), (5) and (6). It would be interesting if further approximations could be constructed, but the labor involved becomes enormous at the next stage.

4. Graphical Representation of a Markoff Process

Stochastic processes of the type described above are known mathematically as discrete Markoff processes and have been extensively studied in the literature.[6] The general case can be described as follows: There exist a finite number of possible "states" of a system; $S_1, S_2, \cdots, S_n$. In addition there is a set of transition probabilities, $p_i(j)$, the probability that if the system is in state S_i it will next go to state S_j. To make this Markoff process into an information source we need only assume that a letter is produced for each transition from one state to another. The states will correspond to the "residue of influence" from preceding letters.

The situation can be represented graphically as shown in Figs. 3, 4 and 5. The "states" are the junction points in the graph and the probabilities and letters produced for a transition are given beside the corresponding line. Figure 3 is for the example B in Section 2, while Fig. 4 corresponds to the example C. In Fig. 3 there is only one state since successive letters are independent. In Fig. 4 there are as many states as letters. If a trigram example were constructed there would be at most n^2 states corresponding to the possible pairs of letters preceding the one being chosen. Figure 5 is a graph for the case of word structure in example D. Here S corresponds to the "space" symbol.

5. Ergodic and Mixed Sources

As we have indicated above a discrete source for our purposes can be considered to be represented by a Markoff process. Among the possible discrete Markoff processes there is a group with special properties of significance in communication theory. This special class consists of the "ergodic" processes and we shall call the corresponding sources ergodic sources. Although a rigorous definition of an ergodic process is somewhat involved, the general idea is simple. In an ergodic process every sequence produced by the process is the same in statistical properties. Thus the letter frequencies, digram frequencies, etc., obtained from particular sequences, will, as the lengths of the sequences increase, approach

[6] For a detailed treatment see M. Frechet, *Methods des fonctions arbitraires. Theorie des énénements en chaine dans le cas d'un nombre fini d'états possibles*. Paris, Gauthier Villars, 1938.

definite limits independent of the particular sequence. Actually this is not true of every sequence but the set for which it is false has probability zero. Roughly the ergodic property means statistical homogeneity.

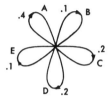

Fig. 3. — A graph corresponding to the source in example B.

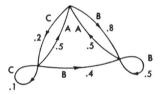

Fig. 4. — A graph corresponding to the source in example C.

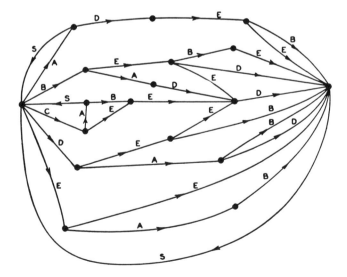

Fig. 5. — A graph corresponding to the source in example D.

All the examples of artificial languages given above are ergodic. This property is related to the structure of the corresponding graph. If the graph has the following two properties[7] the corresponding process will be ergodic:

1. The graph does not consist of two isolated parts A and B such that it is impossible to go from junction points in part A to junction points in part B along lines of the graph in the direction of arrows and also impossible to go from junctions in part B to junctions in part A.

2. A closed series of lines in the graph with all arrows on the lines pointing in the same orientation will be called a "circuit." The "length" of a circuit is the number of lines in it. Thus in Fig. 5 series BEBES is a circuit of length 5. The second property required is that the greatest common divisor of the lengths of all circuits in the graph be one.

If the first condition is satisfied but the second one violated by having the greatest common divisor equal to $d > 1$, the sequences have a certain type of periodic structure. The various sequences fall into d different classes which are statistically the same apart from a shift of the origin (i.e., which letter in the sequence is called letter 1). By a shift of from 0 up to $d - 1$ any sequence can be made statistically equivalent to any other. A simple example with $d = 2$ is the following: There are three possible letters a, b, c. Letter a is followed with either b or c with probabilities $\frac{1}{3}$ and $\frac{2}{3}$ respectively. Either b or c is always followed by letter a. Thus a typical sequence is

$$a\ b\ a\ c\ a\ c\ a\ c\ a\ b\ a\ c\ a\ b\ a\ b\ a\ c\ a\ c.$$

This type of situation is not of much importance for our work.

If the first condition is violated the graph may be separated into a set of subgraphs each of which satisfies the first condition. We will assume that the second condition is also satisfied for each subgraph. We have in this case what may be called a "mixed" source made up of a number of pure components. The components correspond to the various subgraphs. If L_1, L_2, L_3, $\cdots$ are the component sources we may write

---

[7] These are restatements in terms of the graph of conditions given in Frechet.

$$L = p_1 L_1 + p_2 L_2 + p_3 L_3 + \cdots$$

where p_i is the probability of the component source L_i.

Physically the situation represented is this: There are several different sources L_1, L_2, L_3, $\cdots$ which are each of homogeneous statistical structure (i.e., they are ergodic). We do not know *a priori* which is to be used, but once the sequence starts in a given pure component L_i, it continues indefinitely according to the statistical structure of that component.

As an example one may take two of the processes defined above and assume $p_1 = .2$ and $p_2 = .8$. A sequence from the mixed source

$$L = .2L_1 + .8L_2$$

would be obtained by choosing first L_1 or L_2 with probabilities .2 and .8 and after this choice generating a sequence from whichever was chosen.

Except when the contrary is stated we shall assume a source to be ergodic. This assumption enables one to identify averages along a sequence with averages over the ensemble of possible sequences (the probability of a discrepancy being zero). For example the relative frequency of the letter A in a particular infinite sequence will be, with probability one, equal to its relative frequency in the ensemble of sequences.

If P_i is the probability of state i and $p_i(j)$ the transition probability to state j, then for the process to be stationary it is clear that the P_i must satisfy equilibrium conditions:

$$P_j = \sum_i P_i p_i(j).$$

In the ergodic case it can be shown that with any starting conditions the probabilities $P_j(N)$ of being in state j after N symbols, approach the equilibrium values as $N \to \infty$.

6. Choice, Uncertainty and Entropy

We have represented a discrete information source as a Markoff process. Can we define a quantity which will measure, in some sense, how much information is "produced" by such a process, or better, at what rate information is produced?

Suppose we have a set of possible events whose probabilities of occurrence are $p_1, p_2, \cdots, p_n$. These probabilities are known but that is all we know concerning which event will occur. Can we find a measure of how much "choice" is involved in the selection of the event or of how uncertain we are of the outcome?

If there is such a measure, say $H(p_1, p_2, \cdots, p_n)$, it is reasonable to require of it the following properties:

1. H should be continuous in the p_i.

2. If all the p_i are equal, $p_i = \dfrac{1}{n}$, then H should be a monotonic increasing function of n. With equally likely events there is more choice, or uncertainty, when there are more possible events.

3. If a choice be broken down into two successive choices, the original H should be the weighted sum of the individual values of H. The meaning of this is illustrated in Fig. 6. At the left we have three possibilities $p_1 = \frac{1}{2}$, $p_2 = \frac{1}{3}$, $p_3 = \frac{1}{6}$. On the right we first choose between two possibilities each with probability $\frac{1}{2}$, and if the second occurs make another choice with probabilities $\frac{2}{3}$, $\frac{1}{3}$. The final results have the same probabilities as before. We require, in this special case, that

$$H(\tfrac{1}{2}, \tfrac{1}{3}, \tfrac{1}{6}) = H(\tfrac{1}{2}, \tfrac{1}{2}) + \tfrac{1}{2} H(\tfrac{2}{3}, \tfrac{1}{3}).$$

The coefficient $\frac{1}{2}$ is the weighting factor introduced because this second choice only occurs half the time.

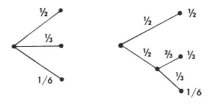

Fig. 6. — Decomposition of a choice from three possibilities.

In Appendix 2, the following result is established:

Theorem 2: The only H satisfying the three above assumptions is of the form:

$$H = -K\sum_{i=1}^{n} p_i \log p_i$$

where K is a positive constant.

This theorem, and the assumptions required for its proof, are in no way necessary for the present theory. It is given chiefly to lend a certain plausibility to some of our later definitions. The real justification of these definitions, however, will reside in their implications.

Quantities of the form $H = -\Sigma\ p_i \log p_i$ (the constant K merely amounts to a choice of a unit of measure) play a central role in information theory as measures of information, choice and uncertainty. The form of H will be recognized as that of entropy

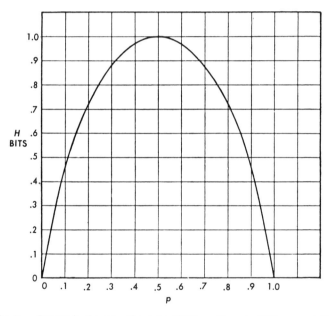

Fig. 7. — Entropy in the case of two possibilities with probabilities p and (1—p).

as defined in certain formulations of statistical mechanics[8] where p_i is the probability of a system being in cell i of its phase space.

[8] See, for example, R. C. Tolman, *Principles of Statistical Mechanics,* Oxford, Clarendon, 1938.

H is then, for example, the H in Boltzmann's famous H theorem. We shall call $H = -\Sigma\, p_i \log p_i$ the entropy of the set of probabilities $p_1, \cdots, p_n$. If x is a chance variable we will write $H(x)$ for its entropy; thus x is not an argument of a function but a label for a number, to differentiate it from $H(y)$ say, the entropy of the chance variable y.

The entropy in the case of two possibilities with probabilities p and $q = 1 - p$, namely

$$H = -(p \log p + q \log q)$$

is plotted in Fig. 7 as a function of p.

The quantity H has a number of interesting properties which further substantiate it as a reasonable measure of choice or information.

1. $H = 0$ if and only if all the p_i but one are zero, this one having the value unity. Thus only when we are certain of the outcome does H vanish. Otherwise H is positive.

2. For a given n, H is a maximum and equal to $\log n$ when all the p_i are equal, i.e., $\frac{1}{n}$. This is also intuitively the most uncertain situation.

3. Suppose there are two events, x and y, in question, with m possibilities for the first and n for the second. Let $p(i, j)$ be the probability of the joint occurrence of i for the first and j for the second. The entropy of the joint event is

$$H(x, y) = -\sum_{i,j} p(i, j) \log p(i, j)$$

while

$$H(x) = -\sum_{i,j} p(i, j) \log \sum_{j} p(i, j)$$
$$H(y) = -\sum_{i,j} p(i, j) \log \sum_{i} p(i, j).$$

It is easily shown that

$$H(x, y) \leq H(x) + H(y)$$

with equality only if the events are independent (i.e., $p(i, j) = p(i)\, p(j)$). The uncertainty of a joint event is less than or equal to the sum of the individual uncertainties.

4. Any change toward equalization of the probabilities p_1, p_2,

$\cdots$, p_n increases H. Thus if $p_1 < p_2$ and we increase p_1, decreasing p_2 an equal amount so that p_1 and p_2 are more nearly equal, then H increases. More generally, if we perform any "averaging" operation on the p_i of the form

$$p_i' = \sum_j a_{ij} p_j$$

where $\sum_i a_{ij} = \sum_j a_{ij} = 1$, and all $a_{ij} \geq 0$, then H increases (except in the special case where this transformation amounts to no more than a permutation of the p_j with H of course remaining the same).

5. Suppose there are two chance events x and y as in 3, not necessarily independent. For any particular value i that x can assume there is a conditional probability $p_i(j)$ that y has the value j. This is given by

$$p_i(j) = \frac{p(i,j)}{\sum_j p(i,j)}.$$

We define the *conditional entropy* of y, $H_x(y)$ as the average of the entropy of y for each value of x, weighted according to the probability of getting that particular x. That is

$$H_x(y) = -\sum_{i,j} p(i,j) \log p_i(j).$$

This quantity measures how uncertain we are of y on the average when we know x. Substituting the value of $p_i(j)$ we obtain

$$H_x(y) = -\sum_{i,j} p(i,j) \log p(i,j) + \sum_{i,j} p(i,j) \log \sum_j p(i,j)$$
$$= H(x,y) - H(x)$$

or

$$H(x,y) = H(x) + H_x(y).$$

The uncertainty (or entropy) of the joint event x, y is the uncertainty of x plus the uncertainty of y when x is known.

6. From 3 and 5 we have

$$H(x) + H(y) \geq H(x,y) = H(x) + H_x(y).$$

Hence

$$H(y) \geq H_x(y).$$

The uncertainty of y is never increased by knowledge of x. It will be decreased unless x and y are independent events, in which case it is not changed.

7. The Entropy of an Information Source

Consider a discrete source of the finite state type considered above. For each possible state i there will be a set of probabilities $p_i(j)$ of producing the various possible symbols j. Thus there is an entropy H_i for each state. The entropy of the source will be defined as the average of these H_i weighted in accordance with the probability of occurrence of the states in question:

$$H = \sum_i P_i H_i$$
$$= -\sum_{i,j} P_i p_i(j) \log p_i(j).$$

This is the entropy of the source per symbol of text. If the Markoff process is proceeding at a definite time rate there is also an entropy per second.

$$H' = \sum_i f_i H_i$$

were f_i is the average frequency (occurrences per second) of state i. Clearly

$$H' = mH$$

where m is the average number of symbols produced per second. H or H' measures the amount of information generated by the source per symbol or per second. If the logarithmic base is 2, they will represent bits per symbol or per second.

If successive symbols are independent then H is simply $-\Sigma\, p_i \log p_i$ where p_i is the probability of symbol i. Suppose in this case we consider a long message of N symbols. It will contain with high probability about $p_1 N$ occurrences of the first symbol, $p_2 N$ occurrences of the second, etc. Hence the probability of this particular message will be roughly

$$p = p_1^{p_1 N}\, p_2^{p_2 N} \cdots p_n^{p_n N}$$

or

$$\log p \doteq N \sum_i p_i \log p_i$$

$$\log p \doteq -NH$$

$$H \doteq \frac{\log 1/p}{N}.$$

H is thus approximately the logarithm of the reciprocal probability of a typical long sequence divided by the number of symbols in the sequence. The same result holds for any source. Stated more precisely we have (see Appendix 3):

Theorem 3: Given any $\epsilon > 0$ and $\delta > 0$, we can find an N_0 such that the sequences of any length $N \geq N_0$ fall into two classes:

1. A set whose total probability is less than ϵ.

2. The remainder, all of whose members have probabilities satisfying the inequality

$$\left| \frac{\log p^{-1}}{N} - H \right| < \delta.$$

In other words we are almost certain to have $\dfrac{\log p^{-1}}{N}$ very close to H when N is large.

A closely related result deals with the number of sequences of various probabilities. Consider again the sequences of length N and let them be arranged in order of decreasing probability. We define $n(q)$ to be the number we must take from this set starting with the most probable one in order to accumulate a total probability q for those taken.

Theorem 4:

$$\lim_{N \to \infty} \frac{\log n(q)}{N} = H$$

when q does not equal 0 or 1.

We may interpret $\log n(q)$ as the number of bits required to specify the sequence when we consider only the most probable sequences with a total probability q. Then $\dfrac{\log n(q)}{N}$ is the number of bits per symbol for the specification. The theorem says that for large N this will be independent of q and equal to H. The rate of growth of the logarithm of the number of reasonably probable sequences is given by H, regardless of our interpretation of "reasonably probable." Due to these results, which are proved in Appendix 3, it is possible for most purposes to treat the long sequences as though there were just 2^{HN} of them, each with a probability 2^{-HN}.

The next two theorems show that H and H' can be determined by limiting operations directly from the statistics of the message sequences, without reference to the states and transition probabilities between states.

Theorem 5: Let $p(B_i)$ be the probability of a sequence B_i of symbols from the source. Let

$$G_N = - \frac{1}{N} \sum_i p(B_i) \log p(B_i)$$

where the sum is over all sequences B_i containing N symbols. Then G_N is a monotonic decreasing function of N and

$$\lim_{N \to \infty} G_N = H.$$

Theorem 6: Let $p(B_i, S_j)$ be the probability of sequence B_i followed by symbol S_j and $p_{B_i}(S_j) = p(B_i, S_j)/p(B_i)$ be the conditional probability of S_j after B_i. Let·

$$F_N = - \sum_{i,j} p(B_i, S_j) \log p_{B_i}(S_j)$$

where the sum is over all blocks B_i of $N - 1$ symbols and over all symbols S_j. Then F_N is a monotonic decreasing function of N,

$$F_N = N G_N - (N - 1) G_{N-1},$$

$$G_N = \frac{1}{N} \sum_1^N F_N,$$

$$F_N \leq G_N,$$

and $\lim_{N \to \infty} F_N = H.$

These results are derived in Appendix 3. They show that a series of approximations to H can be obtained by considering only the statistical structure of the sequences extending over 1, 2, $\cdots$, N symbols. F_N is the better approximation. In fact F_N is the entropy of the Nth order approximation to the source of the type discussed above. If there are no statistical influences extending over more than N symbols, that is if the conditional probability of the next symbol knowing the preceding $(N - 1)$ is not changed by a knowledge of any before that, then $F_N = H$. F_N of course is the conditional entropy of the next symbol when the $(N - 1)$ preceding ones are known, while G_N is the entropy per symbol of blocks of N symbols.

The ratio of the entropy of a source to the maximum value it could have while still restricted to the same symbols will be called its *relative entropy*. This, as will appear later, is the maximum compression possible when we encode into the same alphabet. One minus the relative entropy is the *redundancy*. The redundancy of ordinary English, not considering statistical structure over greater distances than about eight letters, is roughly 50%. This means that when we write English half of what we write is determined by the structure of the language and half is chosen freely. The figure 50% was found by several independent methods which all gave results in this neighborhood. One is by calculation of the entropy of the approximations to English. A second method is to delete a certain fraction of the letters from a sample of English text and then let someone attempt to restore them. If they can be restored when 50% are deleted the redundancy must be greater than 50%. A third method depends on certain known results in cryptography.

Two extremes of redundancy in English prose are represented by Basic English and by James Joyce's book *Finnegans Wake*. The Basic English vocabulary is limited to 850 words and the redundancy is very high. This is reflected in the expansion that occurs when a passage is translated into Basic English. Joyce on the other hand enlarges the vocabulary and is alleged to achieve a compression of semantic content.

The redundancy of a language is related to the existence of crossword puzzles. If the redundancy is zero any sequence of letters is a reasonable text in the language and any two-dimensional array of letters forms a crossword puzzle. If the redundancy is too high the language imposes too many constraints for large crossword puzzles to be possible. A more detailed analysis shows that if we assume the constraints imposed by the language are of a rather chaotic and random nature, large crossword puzzles are just possible when the redundancy is 50%. If the redundancy is 33%, three-dimensional crossword puzzles should be possible, etc.

8. Representation of the Encoding and Decoding Operations

We have yet to represent mathematically the operations performed by the transmitter and receiver in encoding and decoding the information. Either of these will be called a discrete transducer. The input to the transducer is a sequence of input symbols and its output a sequence of output symbols. The transducer may have an internal memory so that its output depends not only on the present input symbol but also on the past history. We assume that the internal memory is finite, i.e., there exist a finite number m of possible states of the transducer and that its output is a function of the present state and the present input symbol. The next state will be a second function of these two quantities. Thus a transducer can be described by two functions:

$$y_n = f(x_n, \alpha_n)$$
$$\alpha_{n+1} = g(x_n, \alpha_n)$$

where:

x_n is the nth input symbol.

α_n is the state of the transducer when the nth input symbol is introduced,

y_n is the output symbol (or sequence of output symbols) produced when x_n is introduced if the state is α_n.

If the output symbols of one transducer can be identified with the input symbols of a second, they can be connected in tandem and the result is also a transducer. If there exists a second transducer which operates on the output of the first and recovers the original input, the first transducer will be called non-singular and the second will be called its inverse.

Theorem 7: The output of a finite state transducer driven by a finite state statistical source is a finite state statistical source, with entropy (per unit time) less than or equal to that of the input. If the transducer is non-singular they are equal.

Let α represent the state of the source, which produces a sequence of symbols x_i; and let β be the state of the transducer, which produces, in its output, blocks of symbols y_j. The combined system can be represented by the "product state space" of pairs (α, β). Two points in the space (α_1, β_1) and (α_2, β_2), are connected by a line if α_1 can produce an x which changes β_1 to β_2, and this line is given the probability of that x in this case. The line is labeled with the block of y_1 symbols produced by the transducer. The entropy of the output can be calculated as the weighted sum over the states. If we sum first on β each resulting term is less than or equal to the corresponding term for α, hence the entropy is not increased. If the transducer is non-singular let its output be connected to the inverse transducer. If H_1', H_2' and H_3' are the output entropies of the source, the first and second transducers respectively, then $H_1' \geq H_2' \geq H_3' = H_1'$ and therefore $H_1' = H_2'$.

Suppose we have a system of constraints on possible sequences of the type which can be represented by a linear graph as in Fig. 2. If probabilities $p_{ij}^{(s)}$ were assigned to the various lines connecting state i to state j this would become a source. There is one particular assignment which maximizes the resulting entropy (see Appendix 4).

Theorem 8: Let the system of constraints considered as a channel have a capacity $C = \log W$. If we assign

$$p_{ij}^{(s)} = \frac{B_j}{B_i} W^{-l_{ij}^{(s)}}$$

where $l_{ij}^{(s)}$ is the duration of the sth symbol leading from state i to state j and the B_i satisfy

$$B_i = \sum_{s,j} B_j W^{-l_{ij}^{(s)}}$$

then H is maximized and equal to C.

By proper assignment of the transition probabilities the entropy of symbols on a channel can be maximized at the channel capacity.

9. The Fundamental Theorem for a Noiseless Channel

We will now justify our interpretation of H as the rate of generating information by proving that H determines the channel capacity required with most efficient coding.

Theorem 9: Let a source have entropy H (bits per symbol) and a channel have a capacity C (bits per second). Then it is possible to encode the output of the source in such a way as to transmit at the average rate $\dfrac{C}{H} - \epsilon$ symbols per second over the channel where ϵ is arbitrarily small. It is not possible to transmit at an average rate greater than $\dfrac{C}{H}$.

The converse part of the theorem, that $\dfrac{C}{H}$ cannot be exceeded, may be proved by noting that the entropy of the channel input per second is equal to that of the source, since the transmitter must be non-singular, and also this entropy cannot exceed the channel capacity. Hence $H' \leq C$ and the number of symbols per second $= H'/H \leq C/H$.

The first part of the theorem will be proved in two different ways. The first method is to consider the set of all sequences of N symbols produced by the source. For N large we can divide these into two groups, one containing less than $2^{(H+\eta)N}$ members and the second containing less than 2^{RN} members (where R is the logarithm of the number of different symbols) and having a total probability less than μ. As N increases η and μ approach zero. The number of signals of duration T in the channel is greater than $2^{(C-\theta)T}$ with θ small when T is large. If we choose

$$T = \left(\frac{H}{C} + \lambda \right) N$$

then there will be a sufficient number of sequences of channel symbols for the high probability group when N and T are sufficiently large (however small λ) and also some additional ones. The high probability group is coded in an arbitrary one-to-one way into this set. The remaining sequences are represented by larger sequences, starting and ending with one of the sequences not used for the high probability group. This special sequence acts as a start and stop signal for a different code. In between a sufficient time is allowed to give enough different sequences for all the low probability messages. This will require

$$T_1 = \left(\frac{R}{C} + \varphi \right) N$$

where φ is small. The mean rate of transmission in message symbols per second will then be greater than

$$\left[(1-\delta) \frac{T}{N} + \delta \frac{T_1}{N} \right]^{-1} = \left[(1-\delta) \left(\frac{H}{C} + \lambda \right) + \delta \left(\frac{R}{C} + \varphi \right) \right]^{-1}.$$

As N increases δ, λ and φ approach zero and the rate approaches $\frac{C}{H}$.

Another method of performing this coding and thereby proving the theorem can be described as follows: Arrange the messages of length N in order of decreasing probability and suppose their probabilities are $p_1 \geq p_2 \geq p_3 \cdot \cdot \cdot \geq p_n$. Let $P_s = \sum_1^{s-1} p_i$; that is P_s is the cumulative probability up to, but not including, p_s. We first encode into a binary system. The binary code for message s is obtained by expanding P_s as a binary number. The expansion is carried out to m_s places, where m_s is the integer satisfying:

$$\log_2 \frac{1}{p_s} \leq m_s < 1 + \log_2 \frac{1}{p_s}.$$

Thus the messages of high probability are represented by short codes and those of low probability by long codes. From these inequalities we have

$$\frac{1}{2^{m_s}} \leq p_s < \frac{1}{2^{m_s-1}}.$$

The code for P_s will differ from all succeeding ones in one or more of its m_s places, since all the remaining P_i are at least $\frac{1}{2^{m_s}}$ larger and their binary expansions therefore differ in the first m_s places. Consequently all the codes are different and it is possible to recover the message from its code. If the channel sequences are not already sequences of binary digits, they can be ascribed binary numbers in an arbitrary fashion and the binary code thus translated into signals suitable for the channel.

The average number H_1 of binary digits used per symbol of original message is easily estimated. We have

$$H_1 = \frac{1}{N} \Sigma m_s p_s.$$

But,

$$\frac{1}{N} \Sigma \left(\log_2 \frac{1}{p_s} \right) p_s \leq \frac{1}{N} \Sigma m_s p_s < \frac{1}{N} \Sigma \left(1 + \log_2 \frac{1}{p_s} \right) p_s$$

and therefore,

$$G_N \leq H_1 < G_N + \frac{1}{N}.$$

As N increases G_N approaches H, the entropy of the source and H_1 approaches H.

We see from this that the inefficiency in coding, when only a finite delay of N symbols is used, need not be greater than $\frac{1}{N}$ plus the difference between the true entropy H and the entropy G_N calculated for sequences of length N. The per cent excess time needed over the ideal is therefore less than

$$\frac{G_N}{H} + \frac{1}{HN} - 1.$$

This method of encoding is substantially the same as one found independently by R. M. Fano.[9] His method is to arrange the messages of length N in order of decreasing probability. Divide this series into two groups of as nearly equal probability as possible. If the message is in the first group its first binary digit will be 0, otherwise 1. The groups are similarly divided into subsets of nearly equal probability and the particular subset determines the second binary digit. This process is continued until each subset contains only one message. It is easily seen that apart from minor differences (generally in the last digit) this amounts to the same thing as the arithmetic process described above.

10. Discussion and Examples

In order to obtain the maximum power transfer from a generator to a load, a transformer must in general be introduced so that

[9] Technical Report No. 65, The Research Laboratory of Electronics, M.I.T., March 17, 1949.

the generator as seen from the load has the load resistance. The situation here is roughly analogous. The transducer which does the encoding should match the source to the channel in a statistical sense. The source as seen from the channel through the transducer should have the same statistical structure as the source which maximizes the entropy in the channel. The content of Theorem 9 is that, although an exact match is not in general possible, we can approximate it as closely as desired. The ratio of the actual rate of transmission to the capacity C may be called the efficiency of the coding system. This is of course equal to the ratio of the actual entropy of the channel symbols to the maximum possible entropy.

In general, ideal or nearly ideal encoding requires a long delay in the transmitter and receiver. In the noiseless case which we have been considering, the main function of this delay is to allow reasonably good matching of probabilities to corresponding lengths of sequences. With a good code the logarithm of the reciprocal probability of a long message must be proportional to the duration of the corresponding signal, in fact

$$\left| \frac{\log p^{-1}}{T} - C \right|$$

must be small for all but a small fraction of the long messages.

If a source can produce only one particular message its entropy is zero, and no channel is required. For example, a computing machine set up to calculate the successive digits of π produces a definite sequence with no chance element. No channel is required to "transmit" this to another point. One could construct a second machine to compute the same sequence at the point. However, this may be impractical. In such a case we can choose to ignore some or all of the statistical knowledge we have of the source. We might consider the digits of π to be a random sequence in that we construct a system capable of sending any sequence of digits. In a similar way we may choose to use some of our statistical knowledge of English in constructing a code, but not all of it. In such a case we consider the source with the maximum entropy subject to the statistical conditions we wish to retain. The entropy of this source determines the channel capacity which is necessary and sufficient. In the π example the only information retained is that all the digits are chosen from the set $0, 1, \cdots, 9$. In the case of English one might wish to use the statistical saving possible due to letter frequencies, but nothing else. The maximum entropy source is then the first approximation to English and its entropy determines the required channel capacity.

As a simple example of some of these results consider a source which produces a sequence of letters chosen from among A, B, C, D with probabilities $\frac{1}{2}, \frac{1}{4}, \frac{1}{8}, \frac{1}{8}$, successive symbols being chosen independently. We have

$$H = - (\tfrac{1}{2} \log \tfrac{1}{2} + \tfrac{1}{4} \log \tfrac{1}{4} + \tfrac{2}{8} \log \tfrac{1}{8})$$
$$= \tfrac{7}{4} \text{ bits per symbol.}$$

Thus we can approximate a coding system to encode messages from this source into binary digits with an average of $\frac{7}{4}$ binary digit per symbol. In this case we can actually achieve the limiting value by the following code (obtained by the method of the second proof of Theorem **9**):

A	0
B	10
C	110
D	111

The average number of binary digits used in encoding a sequence of N symbols will be

$$N \left(\tfrac{1}{2} \times 1 + \tfrac{1}{4} \times 2 + \tfrac{2}{8} \times 3\right) = \tfrac{7}{4} N.$$

It is easily seen that the binary digits 0, 1 have probabilities $\frac{1}{2}, \frac{1}{2}$ so the H for the coded sequences is one bit per symbol. Since, on the average, we have $\frac{7}{4}$ binary symbols per original letter, the entropies on a time basis are the same. The maximum possible entropy for the original set is $\log 4 = 2$, occurring when A, B, C, D have probabilities $\frac{1}{4}, \frac{1}{4}, \frac{1}{4}, \frac{1}{4}$. Hence the relative entropy is $\frac{7}{8}$. We can translate the binary sequences into the original set of symbols on a two-to-one basis by the following table:

00	A'
01	B'
10	C'
11	D'

This double process then encodes the original message into the same symbols but with an average compression ratio $\frac{7}{8}$.

As a second example consider a source which produces a sequence of A's and B's with probability p for A and q for B. If $p < < q$ we have

$$H = - \log p^p (1 - p)^{1-p}$$
$$= - p \log p (1 - p)^{(1-p)/p}$$
$$= p \log \frac{e}{p}.$$

In such a case one can construct a fairly good coding of the message on a 0, 1 channel by sending a special sequence, say 0000, for the infrequent symbol A and then a sequence indicating the *number* of B's following it. This could be indicated by the binary representation with all numbers containing the special sequence deleted. All numbers up to 16 are represented as usual; 16 is represented by the next binary number after 16 which does not contain four zeros, namely $17 = 10001$, etc.

It can be shown that as $p \to 0$ the coding approaches ideal provided the length of the special sequence is properly adjusted.

II

The Discrete Channel with Noise

11. Representation of a Noisy Discrete Channel

We now consider the case where the signal is perturbed by noise during transmission or at one or the other of the terminals. This means that the received signal is not necessarily the same as that sent out by the transmitter. Two cases may be distinguished. If a particular transmitted signal always produces the same received signal, i.e., the received signal is a definite function of the transmitted signal, then the effect may be called distortion. If this function has an inverse — no two transmitted signals producing the same received signal — distortion may be corrected, at least in principle, by merely performing the inverse functional operation on the received signal.

The case of interest here is that in which the signal does not always undergo the same change in transmission. In this case we may assume the received signal E to be a function of the transmitted signal S and a second variable, the noise N.

$$E = f(S,N)$$

The noise is considered to be a chance variable just as the message was above. In general it may be represented by a suitable stochastic process. The most general type of noisy discrete channel we shall consider is a generalization of the finite state noise-free channel described previously. We assume a finite number of states and a set of probabilities

$$p_{\alpha,i}(\beta,j).$$

This is the probability, if the channel is in state α and symbol i is transmitted, that symbol j will be received and the channel left in state β. Thus α and β range over the possible states, i over the possible transmitted signals and j over the possible received signals. In the case where successive symbols are independently perturbed by the noise there is only one state, and the channel is described by the set of transition probabilities $p_i(j)$, the probability of transmitted symbol i being received as j.

If a noisy channel is fed by a source there are two statistical processes at work: the source and the noise. Thus there are a number of entropies that can be calculated. First there is the entropy $H(x)$ of the source or of the input to the channel (these will be equal if the transmitter is non-singular). The entropy of the output of the channel, i.e., the received signal, will be denoted by $H(y)$. In the noiseless case $H(y) = H(x)$. The joint entropy of input and output will be $H(x,y)$. Finally there are two conditional entropies $H_x(y)$ and $H_y(x)$, the entropy of the output when the input is known and conversely. Among these quantities we have the relations

$$H(x,y) = H(x) + H_x(y) = H(y) + H_y(x).$$

All of these entropies can be measured on a per-second or a per-symbol basis.

12. Equivocation and Channel Capacity

If the channel is noisy it is not in general possible to reconstruct the original message or the transmitted signal with *certainty* by any operation on the received signal E. There are, however, ways of transmitting the information which are optimal in combating noise. This is the problem which we now consider.

Suppose there are two possible symbols 0 and 1, and we are transmitting at a rate of 1000 symbols per second with probabilities $p_0 = p_1 = \frac{1}{2}$. Thus our source is producing information at the rate of 1000 bits per second. During transmission the noise introduces errors so that, on the average, 1 in 100 is received incorrectly (a 0 as 1, or 1 as 0). What is the rate of transmission of information? Certainly less than 1000 bits per second since about 1% of the received symbols are incorrect. Our first impulse might be to say the rate is 990 bits per second, merely subtracting the expected number of errors. This is not satisfactory since it fails to take into account the recipient's lack of knowledge of where the errors occur. We may carry it to an extreme case and suppose the noise so great that the received symbols are entirely independent of the transmitted symbols. The probability of receiving 1 is $\frac{1}{2}$ whatever was transmitted and similarly for 0. Then about half of the received symbols are correct due to chance alone, and we would be giving the system credit for transmitting 500 bits per second while actually no information is being transmitted at all. Equally "good" transmission would be obtained by dispensing with the channel entirely and flipping a coin at the receiving point.

Evidently the proper correction to apply to the amount of information transmitted is the amount of this information which is missing in the received signal, or alternatively the uncertainty when we have received a signal of what was actually sent. From our previous discussion of entropy as a measure of uncertainty it seems reasonable to use the conditional entropy of the message, knowing the received signal, as a measure of this missing information. This is indeed the proper definition, as we shall see later. Following this idea the rate of actual transmission, R, would be obtained by subtracting from the rate of production (i.e., the entropy of the source) the average rate of conditional entropy.

$$R = H(x) - H_y(x)$$

The conditional entropy $H_y(x)$ will, for convenience, be called the equivocation. It measures the average ambiguity of the received signal.

In the example considered above, if a 0 is received the *a posteriori* probability that a 0 was transmitted is .99, and that a 1 was transmitted is .01. These figures are reversed if a 1 is received. Hence

$$H_y(x) = - [.99 \log .99 + 0.01 \log 0.01]$$
$$= .081 \text{ bits/symbol}$$

or 81 bits per second. We may say that the system is transmitting at a rate $1000 - 81 = 919$ bits per second. In the extreme case where a 0 is equally likely to be received as a 0 or 1 and similarly for 1, the *a posteriori* probabilities are $\frac{1}{2}$, $\frac{1}{2}$ and

$$H_y(x) = -\left[\tfrac{1}{2} \log \tfrac{1}{2} + \tfrac{1}{2} \log \tfrac{1}{2}\right]$$
$$= 1 \text{ bit per symbol}$$

or 1000 bits per second. The rate of transmission is then 0 as it should be.

The following theorem gives a direct intuitive interpretation of the equivocation and also serves to justify it as the unique appropriate measure. We consider a communication system and an observer (or auxiliary device) who can see both what is sent and what is recovered (with errors due to noise). This observer notes the errors in the recovered message and transmits data to the receiving point over a "correction channel" to enable the receiver to correct the errors. The situation is indicated schematically in Fig. 8.

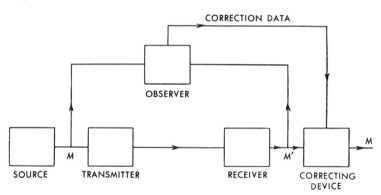

Fig. 8. — Schematic diagram of a correction system.

Theorem 10: If the correction channel has a capacity equal to $H_y(x)$ it is possible to so encode the correction data as to send it over this channel and correct all but an arbitrarily small fraction ϵ of the errors. This is not possible if the channel capacity is less than $H_y(x)$.

Roughly then, $H_y(x)$ is the amount of additional information that must be supplied per second at the receiving point to correct the received message.

To prove the first part, consider long sequences of received message M' and corresponding original message M. There will be logarithmically $TH_y\,(x)$ of the M's which could reasonably have produced each M'. Thus we have $TH_y(x)$ binary digits to send each T seconds. This can be done with ϵ frequency of errors on a channel of capacity $H_y(x)$.

The second part can be proved by noting, first, that for any discrete chance variables x, y, z

$$H_y(x, z) \geq H_y(x).$$

The left-hand side can be expanded to give

$$H_y(z) + H_{yz}(x) \geq H_y(x)$$
$$H_{yz}(x) \geq H_y(x) - H_y(z) \geq H_y(x) - H(z).$$

If we identify x as the output of the source, y as the received signal and z as the signal sent over the correction channel, then the right-hand side is the equivocation less the rate of transmission over the correction channel. If the capacity of this channel is less than the equivocation the right-hand side will be greater than zero and $H_{yz}(x) > 0$. But this is the uncertainty of what was sent, knowing both the received signal and the correction signal. If this is greater than zero the frequency of errors cannot be arbitrarily small.

Example:

> Suppose the errors occur at random in a sequence of binary digits: probability p that a digit is wrong and $q = 1 - p$ that it is right. These errors can be corrected if their position is known. Thus the correction channel need only send information as to these positions. This amounts to transmitting from a source which produces binary digits with probability p for 1 (incorrect) and q for 0 (correct). This requires a channel of capacity
>
> $$-[p \log p + q \log q]$$
>
> which is the equivocation of the original system.

The rate of transmission R can be written in two other forms due to the identities noted above. We have

$$R = H(x) - H_y(x)$$
$$= H(y) - H_x(y)$$
$$= H(x) + H(y) - H(x, y).$$

The first defining expression has already been interpreted as the amount of information sent less the uncertainty of what was sent. The second measures the amount received less the part of this which is due to noise. The third is the sum of the two amounts less the joint entropy and therefore in a sense is the number of bits per second common to the two. Thus all three expressions have a certain intuitive significance.

The capacity C of a noisy channel should be the maximum possible rate of transmission, i.e., the rate when the source is properly matched to the channel. We therefore define the channel capacity by

$$C = \text{Max} \ (H(x) - H_y(x))$$

where the maximum is with respect to all possible information sources used as input to the channel. If the channel is noiseless, $H_y(x) = 0$. The definition is then equivalent to that already given for a noiseless channel since the maximum entropy for the channel is its capacity by Theorem 8.

13. The Fundamental Theorem for a Discrete Channel with Noise

It may seem surprising that we should define a definite capacity C for a noisy channel since we can never send certain information in such a case. It is clear, however, that by sending the information in a redundant form the probability of errors can be reduced. For example, by repeating the message many times and by a statistical study of the different received versions of the message the probability of errors could be made very small. One would expect, however, that to make this probability of errors approach zero, the redundancy of the encoding must increase indefinitely, and the rate of transmission therefore approach zero. This is by no means true. If it were, there would not be a very well defined capacity, but only a capacity for a given frequency of errors, or a given equivocation; the capacity going down as the error requirements are made more stringent. Actually the capacity C defined above has a very definite significance. It is possible to send information at the rate C through the channel *with as small a frequency of errors or equivocation as desired* by proper encoding. This statement is not true for any rate greater than C. If an attempt is made to transmit at a higher rate than C, say $C + R_1$, then there will necessarily be an equivocation equal to or greater than the excess R_1. Nature takes payment by requiring just that much uncertainty, so that we are not actually getting any more than C through correctly.

The situation is indicated in Fig. 9. The rate of information into the channel is plotted horizontally and the equivocation vertically. Any point above the heavy line in the shaded region can be attained and those below cannot. The points on the line cannot in general be attained, but there will usually be two points on the line that can.

These results are the main justification for the definition of C and will now be proved.

Theorem 11: Let a discrete channel have the capacity C and a discrete source the entropy per second H. If $H \leq C$ there exists a coding system such that the output of the source can be transmitted over the channel with an arbitrarily small frequency of errors (or an arbitrarily small equivocation). If $H > C$ it is possible to encode the source so that the equivocation is less than $H - C + \epsilon$ where ϵ is arbitrarily small. There is no method of encoding which gives an equivocation less than $H - C$.

The method of proving the first part of this theorem is not by exhibiting a coding method having the desired properties, but by showing that such a code must exist in a certain group of codes.

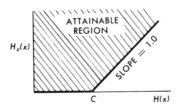

Fig. 9. — The equivocation possible for a given input entropy to a channel.

In fact we will average the frequency of errors over this group and show that this average can be made less than ϵ. If the average of a set of numbers is less than ϵ there must exist at least one in the set which is less than ϵ. This will establish the desired result.

The capacity C of a noisy channel has been defined as

$$C = \text{Max} \, (H(x) - H_y(x))$$

where x is the input and y the output. The maximization is over all sources which might be used as input to the channel.

Let S_0 be a source which achieves the maximum capacity C. If this maximum is not actually achieved by any source (but only approached as a limit) let S_0 be a source which approximates to giving the maximum rate. Suppose S_0 is used as input to the channel. We consider the possible transmitted and received sequences of a long duration T. The following will be true:

1. The transmitted sequences fall into two classes, a high probability group with about $2^{TH(x)}$ members and the remaining sequences of small total probability.

2. Similarly the received sequences have a high probability set of about $2^{TH(y)}$ members and a low probability set of remaining sequences.

3. Each high probability output could be produced by about $2^{TH_y(x)}$ inputs. The total probability of all other cases is small.

4. Each high probability input could result in about $2^{TH_x(y)}$ outputs. The total probability of all other results is small.

All the ϵ's and δ's implied by the words "small" and "about" in these statements approach zero as we allow T to increase and S_0 to approach the maximizing source.

The situation is summarized in Fig. 10 where the input sequences are points on the left and output sequences points on the right. The upper fan of cross lines represents the range of possible causes for a typical output. The lower fan represents the range of possible results from a typical input. In both cases the "small probability" sets are ignored.

Now suppose we have another source S, producing information at rate R with $R < C$. In the period T this source will have 2^{TR} high probability messages. We wish to associate these with a

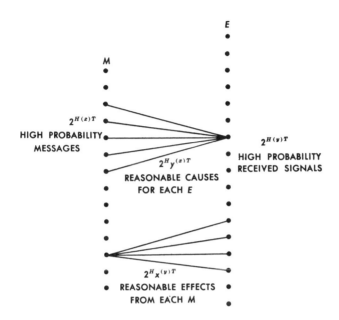

Fig. 10. — Schematic representation of the relations between inputs and outputs in a channel.

selection of the possible channel inputs in such a way as to get a small frequency of errors. We will set up this association in all possible ways (using, however, only the high probability group of inputs as determined by the source S_0) and average the frequency of errors for this large class of possible coding systems. This is the same as calculating the frequency of errors for a random association of the messages and channel inputs of duration T. Suppose a particular output y_1 is observed. What is the probability of more than one message from S in the set of possible causes of y_1? There are 2^{TR} messages distributed at random in $2^{TH(x)}$ points. The probability of a particular point being a message is thus

$$2^{T(R-H(x))}.$$

The probability that none of the points in the fan is a message (apart from the actual originating message) is

$$P = \left[1 - 2^{T(R-H(x))}\right]^{2^{TH_y(x)}}.$$

Now $R < H(x) - H_y(x)$ so $R - H(x) = -H_y(x) - \eta$ with η positive. Consequently

$$P = [1 - 2^{-TH_y(x)-T\eta}]^{2^{TH_y(x)}}$$

approaches (as $T \to \infty$)

$$1 - 2^{-T\eta}.$$

Hence the probability of an error approaches zero and the first part of the theorem is proved.

The second part of the theorem is easily shown by noting that we could merely send C bits per second from the source, completely neglecting the remainder of the information generated. At the receiver the neglected part gives an equivocation $H(x) - C$ and the part transmitted need only add ϵ. This limit can also be attained in many other ways, as will be shown when we consider the continuous case.

The last statement of the theorem is a simple consequence of our definition of C. Suppose we can encode a source with $H(x) = C + a$ in such a way as to obtain an equivocation $H_y(x) = a - \epsilon$ with ϵ positive. Then

$$H(x) - H_y(x) = C + \epsilon$$

with ϵ positive. This contradicts the definition of C as the maximum of $H(x) - H_y(x)$.

Actually more has been proved than was stated in the theorem. If the average of a set of positive numbers is within ϵ of zero, a fraction of at most $\sqrt{\epsilon}$ can have values greater than $\sqrt{\epsilon}$. Since ϵ is arbitrarily small we can say that almost all the systems are arbitrarily close to the ideal.

14. Discussion

The demonstration of Theorem 11, while not a pure existence proof, has some of the deficiencies of such proofs. An attempt to obtain a good approximation to ideal coding by following the method of the proof is generally impractical. In fact, apart from some rather trivial cases and certain limiting situations, no explicit description of a series of approximation to the ideal has been found. Probably this is no accident but is related to the difficulty of giving an explicit construction for a good approximation to a random sequence.

An approximation to the ideal would have the property that if the signal is altered in a reasonable way by the noise, the original can still be recovered. In other words the alteration will not in general bring it closer to another reasonable signal than the original. This is accomplished at the cost of a certain amount of redundancy in the coding. The redundancy must be introduced in the proper way to combat the particular noise structure involved. However, any redundancy in the source will usually help if it is utilized at the receiving point. In particular, if the source already has a certain redundancy and no attempt is made to eliminate it in matching to the channel, this redundancy will help combat noise. For example, in a noiseless telegraph channel one could save about 50% in time by proper encoding of the messages. This is not done and most of the redundancy of English remains in the channel symbols. This has the advantage, however, of allowing considerable noise in the channel. A sizable fraction of the letters can be received incorrectly and still reconstructed by the context. In fact this is probably not a bad approximation to the ideal in many cases, since the statistical structure of English is rather involved and the reasonable English sequences are not too far (in the sense required for the theorem) from a random selection.

As in the noiseless case a delay is generally required to approach the ideal encoding. It now has the additional function of allowing a large sample of noise to affect the signal before any judgment is made at the receiving point as to the original message. Increasing the sample size always sharpens the possible statistical assertions.

The content of Theorem 11 and its proof can be formulated in a somewhat different way which exhibits the connection with the noiseless case more clearly. Consider the possible signals of duration T and suppose a subset of them is selected to be used. Let those in the subset all be used with equal probability, and suppose the receiver is constructed to select, as the original signal, the most probable cause from the subset, when a perturbed signal is received. We define $N(T, q)$ to be the maximum number of signals we can choose for the subset such that the probability of an incorrect interpretation is less than or equal to q.

Theorem 12: $\underset{T\to\infty}{\mathrm{Lim}}\ \dfrac{\log N(T, q)}{T} = C$, *where C is the channel capacity, provided that q does not equal 0 or 1.*

In other words, no matter how we set our limits of reliability, we can distinguish reliably in time T enough messages to correspond to about CT bits, when T is sufficiently large. Theorem 12 can be compared with the definition of the capacity of a noiseless channel given in section 1.

15. Example of a Discrete Channel and Its Capacity

A simple example of a discrete channel is indicated in Fig. 11. There are three possible symbols. The first is never affected by noise. The second and third each have probability p of coming through undisturbed, and q of being changed into the other of the

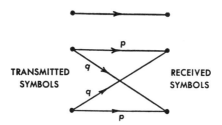

Fig. 11. — Example of a discrete channel.

pair. Let $\alpha = - [p \log p + q \log q]$ and let P, Q and Q be the probabilities of using the first, second and third symbols respectively (the last two being equal from consideration of symmetry). We have:

$$H(x) = -P \log P - 2Q \log Q$$
$$H_y(x) = 2Q\alpha.$$

We wish to choose P and Q in such a way as to maximize $H(x) - H_y(x)$, subject to the constraint $P + 2Q = 1$. Hence we consider

$$U = -P \log P - 2Q \log Q - 2Q\alpha + \lambda(P + 2Q)$$

$$\frac{\partial U}{\partial P} = -1 - \log P + \lambda = 0$$

$$\frac{\partial U}{\partial Q} = -2 - 2 \log Q - 2\alpha + 2\lambda = 0.$$

Eliminating λ

$$\log P = \log Q + \alpha$$
$$P = Qe^\alpha = Q\beta$$
$$P = \frac{\beta}{\beta + 2} \qquad Q = \frac{1}{\beta + 2}.$$

The channel capacity is then

$$C = \log \frac{\beta + 2}{\beta}.$$

Note how this checks the obvious values in the cases $p = 1$ and $p = \frac{1}{2}$. In the first, $\beta = 1$ and $C = \log 3$, which is correct since the channel is then noiseless with three possible symbols. If $p = \frac{1}{2}$, $\beta = 2$ and $C = \log 2$. Here the second and third symbols cannot be distinguished at all and act together like one symbol. The first symbol is used with probability $P = \frac{1}{2}$ and the second and third together with probability $\frac{1}{2}$. This may be distributed between them in any desired way and still achieve the maximum capacity.

For intermediate values of p the channel capacity will lie between $\log 2$ and $\log 3$. The distinction between the second and third symbols conveys some information but not as much as in the noiseless case. The first symbol is used somewhat more frequently than the other two because of its freedom from noise.

16. The Channel Capacity in Certain Special Cases

If the noise affects successive channel symbols independently it can be described by a set of transition probabilities p_{ij}. This is the probability, if symbol i is sent, that j will be received. The channel capacity is then given by the maximum of

$$-\sum_{i,j} P_i p_{ij} \log \sum_i P_i p_{ij} + \sum_{i,j} P_i p_{ij} \log p_{ij}$$

where we vary the P_i subject to $\Sigma P_i = 1$. This leads by the method of Lagrange to the equations,

$$\sum_j p_{sj} \log \frac{p_{sj}}{\sum_i P_i p_{ij}} = \mu \qquad s = 1, 2, \cdots .$$

Multiplying by P_s and summing on s shows that $\mu = C$. Let the inverse of p_{sj} (if it exists) be h_{st} so that $\sum_s h_{st} p_{sj} = \delta_{tj}$. Then:

$$\sum_{s,j} h_{st} p_{sj} \log p_{sj} - \log \sum_i P_i p_{it} = C \sum_s h_{st}.$$

Hence:

$$\sum_i P_i p_{it} = \exp\left[-C \sum_s h_{st} + \sum_{s,j} h_{st} p_{sj} \log p_{sj} \right]$$

or,

$$P_i = \sum_t h_{it} \exp\left[-C \sum_s h_{st} + \sum_{s,j} h_{st} p_{sj} \log p_{sj} \right].$$

This is the system of equations for determining the maximizing values of P_i, with C to be determined so that $\Sigma P_i = 1$. When this is done C will be the channel capacity, and the P_i the proper probabilities for the channel symbols to achieve this capacity.

If each input symbol has the same set of probabilities on the lines emerging from it, and the same is true of each output sym-

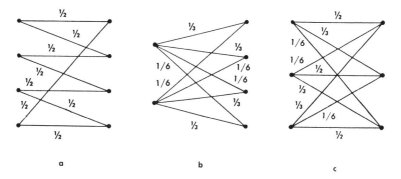

Fig. 12. — Examples of discrete channels with the same transition probabilities for each input and for each output.

bol, the capacity can be easily calculated. Examples are shown in Fig. 12. In such a case $H_x(y)$ is independent of the distribution of probabilities on the input symbols, and is given by $- \Sigma\, p_i \log p_i$ where the p_i are the values of the transition probabilities from any input symbol. The channel capacity is

$$\text{Max}\,[H(y) - H_x(y)]$$
$$= \text{Max}\, H(y) + \Sigma\, p_i \log p_i.$$

The maximum of $H(y)$ is clearly $\log m$ where m is the number of output symbols, since it is possible to make them all equally probable by making the input symbols equally probable. The channel capacity is therefore

$$C = \log m + \Sigma\, p_i \log p_i.$$

In Fig. 12a it would be

$$C = \log 4 - \log 2 = \log 2.$$

This could be achieved by using only the 1st and 3d symbols. In Fig. 12b

$$C = \log 4 - \tfrac{2}{3} \log 3 - \tfrac{1}{3} \log 6$$
$$= \log 4 - \log 3 - \tfrac{1}{3} \log 2$$
$$= \log \tfrac{1}{3}\, 2^{\frac{5}{3}}.$$

In Fig. 12c we have

$$C = \log 3 - \tfrac{1}{2} \log 2 - \tfrac{1}{3} \log 3 - \tfrac{1}{6} \log 6$$
$$= \log \frac{3}{2^{\frac{1}{2}}\, 3^{\frac{1}{3}}\, 6^{\frac{1}{6}}}.$$

Suppose the symbols fall into several groups such that the noise never causes a symbol in one group to be mistaken for a symbol in another group. Let the capacity for the nth group be C_n (in bits per second) when we use only the symbols in this group. Then it is easily shown that, for best use of the entire set, the total probability P_n of all symbols in the nth group should be

$$P_n = \frac{2^{C_n}}{\Sigma 2^{C_n}}.$$

Within a group the probability is distributed just as it would be if these were the only symbols being used. The channel capacity is

$$C = \log \Sigma 2^{C_n}.$$

17. An Example of Efficient Coding

The following example, although somewhat artificial, is a case in which exact matching to a noisy channel is possible. There are two channel symbols, 0 and 1, and the noise affects them in blocks of seven symbols. A block of seven is either transmitted without error, or exactly one symbol of the seven is incorrect. These eight possibilities are equally likely. We have

$$C = \text{Max} \, [H(y) - H_x(y)]$$
$$= \tfrac{1}{7} \, [7 + \tfrac{8}{8} \log \tfrac{1}{8}]$$
$$= \tfrac{4}{7} \text{ bits/symbol.}$$

An efficient code, allowing complete correction of errors and transmitting at the rate C, is the following (found by a method due to R. Hamming):

Let a block of seven symbols be $X_1, X_2, \cdots, X_7$. Of these X_3, X_5, X_6 and X_7 are message symbols and chosen arbitrarily by the source. The other three are redundant and calculated as follows:

X_4 is chosen to make $\alpha = X_4 + X_5 + X_6 + X_7$ even
X_2 "   "   "   " $\beta = X_2 + X_3 + X_6 + X_7$ "
X_1 "   "   "   " $\gamma = X_1 + X_3 + X_5 + X_7$ "

When a block of seven is received α, β and γ are calculated and if even called zero, if odd called one. The binary number $\alpha \, \beta \, \gamma$ then gives the subscript of the X_i that is incorrect (if 0 there was no error).[10]

---

[10] For some further examples of self-correcting codes see M. J. E. Golay, "Notes on Digital Coding," *Proceedings of the Institute of Radio Engineers*, v. 37, No. 6, June, 1949, p. 637.

III
Continuous Information

We now consider the case where the signals or the messages or both are continuously variable, in contrast with the discrete nature assumed heretofore. To a considerable extent the continuous case can be obtained through a limiting process from the discrete case by dividing the continuum of messages and signals into a large but finite number of small regions and calculating the various parameters involved on a discrete basis. As the size of the regions is decreased these parameters in general approach as limits the proper values for the continuous case. There are, however, a few new effects that appear and also a general change of emphasis in the direction of specialization of the general results to particular cases.

We will not attempt, in the continuous case, to obtain our results with the greatest generality, or with the extreme rigor of pure mathematics, since this would involve a great deal of abstract measure theory and would obscure the main thread of the analysis. A preliminary study, however, indicates that the theory can be formulated in a completely axiomatic and rigorous manner which includes both the continuous and discrete cases and many others. The occasional liberties taken with limiting processes in the present analysis can be justified in all cases of practical interest.

18. Sets and Ensembles of Functions

We shall have to deal in the continuous case with sets of functions and ensembles of functions. A set of functions, as the name implies, is merely a class or collection of functions, generally of one variable, time. It can be specified by giving an explicit representation of the various functions in the set, or implicitly by giving a property which functions in the set possess and others do not. Some examples are:

1. The set of functions:

$$f_\theta(t) = \sin (t + \theta).$$

   Each particular value of θ determines a particular function in the set.

2. The set of all functions of time containing no frequencies over W cycles per second.

3. The set of all functions limited in band to W and in amplitude to A.

4. The set of all English speech signals as functions of time.

An *ensemble* of functions is a set of functions together with a probability measure whereby we may determine the probability of a function in the set having certain properties.[1] For example with the set,

$$f_\theta(t) = \sin (t + \theta),$$

we may give a probability distribution for θ, say $P(\theta)$. The set then becomes an ensemble.

Some further examples of ensembles of functions are:

1. A finite set of functions $f_k(t)$ ($k = 1, 2, \cdots, n$) with the probability of f_k being p_k.

2. A finite dimensional family of functions

$$f(\alpha_1, \alpha_2, \cdots, \alpha_n; t)$$

   with a probability distribution for the parameters α_i:

$$p(\alpha_1, \cdots, \alpha_n).$$

   For example we could consider the ensemble defined by

$$f(a_1, \cdots, a_n, \theta_1, \cdots, \theta_n; t) = \sum_{n=1}^{n} a_n \sin n(\omega t + \theta_n)$$

[1] In mathematical terminology the functions belong to a measure space whose total measure is unity.

with the amplitudes a_i distributed normally and independently, and the phases θ_i distributed uniformly (from 0 to 2π) and independently.

3. The ensemble

$$f(a_i, t) = \sum_{n=-\infty}^{+\infty} a_n \frac{\sin \pi(2Wt - n)}{\pi(2Wt - n)}$$

with the a_i normal and independent all with the same standard deviation $\sqrt{N}$. This is a representation of "white" noise, band limited to the band from 0 to W cycles per second and with average power N.[2]

4. Let points be distributed on the t axis according to a Poisson distribution. At each selected point the function $f(t)$ is placed and the different functions added, giving the ensemble

$$\sum_{k=-\infty}^{\infty} f(t + t_k)$$

where the t_k are the points of the Poisson distribution. This ensemble can be considered as a type of impulse or shot noise where all the impulses are identical.

5. The set of English speech functions with the probability measure given by the frequency of occurrence in ordinary use.

An ensemble of functions $f_\alpha(t)$ is *stationary* if the same ensemble results when all functions are shifted any fixed amount in time. The ensemble

$$f_\theta(t) = \sin (t + \theta)$$

is stationary if θ is distributed uniformly from 0 to 2π. If we shift each function by t_1 we obtain

$$f_\theta(t + t_1) = \sin (t + t_1 + \theta)$$
$$= \sin (t + \varphi)$$

---

[2] This representation can be used as a definition of band limited white noise. It has certain advantages in that it involves fewer limiting operations than do definitions that have been used in the past. The name "white noise," already firmly intrenched in the literature, is perhaps somewhat unfortunate. In optics white light means either any continuous spectrum as contrasted with a point spectrum, or a spectrum which is flat with *wavelength* (which is not the same as a spectrum flat with frequency).

with φ distributed uniformly from 0 to 2π. Each function has changed but the ensemble as a whole is invariant under the translation. The other examples given above are also stationary.

An ensemble is *ergodic* if it is stationary, and there is no subset of the functions in the set with a probability different from 0 and 1 which is stationary. The ensemble

$$\sin (t + \theta)$$

is ergodic. No subset of these functions of probability $\neq 0$, 1 is transformed into itself under all time translations. On the other hand the ensemble

$$a \sin (t + \theta)$$

with a distributed normally and θ uniform is stationary but not ergodic. The subset of these functions with a between 0 and 1, for example, is stationary, and has a probability not equal to 0 or 1.

Of the examples given, 3 and 4 are ergodic, and 5 may perhaps be considered so. If an ensemble is ergodic we may say roughly that each function in the set is typical of the ensemble. More precisely it is known that with an ergodic ensemble an average of any statistic over the ensemble is equal (with probability 1) to an average over all the time translations of a particular function in the set.[3] Roughly speaking, each function can be expected, as time progresses, to go through, with the proper frequency, all the convolutions of any of the functions in the set.

Just as we may perform various operations on numbers or functions to obtain new numbers or functions, we can perform operations on ensembles to obtain new ensembles. Suppose, for example, we have an ensemble of functions $f_\alpha(t)$ and an operator T which gives for each function $f_\alpha(t)$ a resulting function $g_\alpha(t)$:

$$g_\alpha(t) = Tf_\alpha(t).$$

[3] This is the famous ergodic theorem or rather one aspect of this theorem which was proved in somewhat different formulations by Birkhoff, von Neumann, and Koopman, and subsequently generalized by Wiener, Hopf, Hurewicz and others. The literature on ergodic theory is quite extensive and the reader is referred to the papers of these writers for precise and general formulations; e.g., E. Hopf "Ergodentheorie," *Ergebnisse der Mathematic und ihrer Grenzgebiete,* v. 5; "On Causality Statistics and Probability," *Journal of Mathematics and Physics,* v. XIII, No. 1, 1934; N. Wiener "The Ergodic Theorem," *Duke Mathematical Journal,* v. 5, 1939.

Probability measure is defined for the set $g_\alpha(t)$ by means of that for the set $f_\alpha(t)$. The probability of a certain subset of the $g_\alpha(t)$ functions is equal to that of the subset of the $f_\alpha(t)$ functions which produce members of the given subset of g functions under the operation T. Physically this corresponds to passing the ensemble through some device, for example, a filter, a rectifier or a modulator. The output functions of the device form the ensemble $g_\alpha(t)$.

A device or operator T will be called invariant if shifting the input merely shifts the output, i.e., if

$$g_\alpha(t) = T f_\alpha(t).$$

implies

$$g_\alpha(t + t_1) = T f_\alpha(t + t_1)$$

for all $f_\alpha(t)$ and all t_1. It is easily shown (see Appendix 5) that if T is invariant and the input ensemble is stationary then the output ensemble is stationary. Likewise if the input is ergodic the output will also be ergodic.

A filter or a rectifier is invariant under all time translations. The operation of modulation is not, since the carrier phase gives a certain time structure. However, modulation is invariant under all translations which are multiples of the period of the carrier.

Wiener has pointed out the intimate relation between the invariance of physical devices under time translations and Fourier theory.[4] He has shown, in fact, that if a device is linear as well as invariant Fourier analysis is then the appropriate mathematical tool for dealing with the problem.

An ensemble of functions is the appropriate mathematical representation of the messages produced by a continuous source (for example, speech), of the signals produced by a transmitter, and of the perturbing noise. Communication theory is properly

[4] Communication theory is heavily indebted to Wiener for much of its basic philosophy and theory. His classic NDRC report, *The Interpolation, Extrapolation, and Smoothing of Stationary Time Series* (Wiley, 1949), contains the first clear-cut formulation of communication theory as a statistical problem, the study of operations on time series. This work, although chiefly concerned with the linear prediction and filtering problem, is an important collateral reference in connection with the present paper. We may also refer here to Wiener's *Cybernetics* (Wiley, 1948), dealing with the general problems of communication and control.

concerned, as has been emphasized by Wiener, not with operations on particular functions, but with operations on ensembles of functions. A communication system is designed not for a particular speech function and still less for a sine wave, but for the ensemble of speech functions.

19. Band Limited Ensembles of Functions

If a function of time $f(t)$ is limited to the band from 0 to W cycles per second it is completely determined by giving its ordinates at a series of discrete points spaced $\dfrac{1}{2W}$ seconds apart in the manner indicated by the following result.[5]

Theorem 13: Let $f(t)$ contain no frequencies over W. Then

$$f(t) = \sum_{-\infty}^{\infty} X_n \frac{\sin \pi(2Wt - n)}{\pi(2Wt - n)}$$

where

$$X_n = f\left(\frac{n}{2W}\right).$$

In this expansion $f(t)$ is represented as a sum of orthogonal functions. The coefficients X_n of the various terms can be considered as coordinates in an infinite dimensional "function space." In this space each function corresponds to precisely one point and each point to one function.

A function can be considered to be substantially limited to a time T if all the ordinates X_n outside this interval of time are zero. In this case all but $2TW$ of the coordinates will be zero. Thus functions limited to a band W and duration T correspond to points in a space of $2TW$ dimensions.

A subset of the functions of band W and duration T corresponds to a region in this space. For example, the functions whose total energy is less than or equal to E correspond to points in a $2TW$ dimensional sphere with radius $r = \sqrt{2WE}$.

An *ensemble* of functions of limited duration and band will be

[5] For a proof of this theorem and further discussion see the author's paper "Communication in the Presence of Noise" published in the *Proceedings of the Institute of Radio Engineers,* v. 37, No. 1, Jan., 1949, pp. 10-21.

represented by a probability distribution $p(x_1, \cdot \cdot \cdot, x_n)$ in the corresponding n dimensional space. If the ensemble is not limited in time we can consider the $2TW$ coordinates in a given interval T to represent substantially the part of the function in the interval T and the probability distribution $p(x_1, \cdot \cdot \cdot, x_n)$ to give the statistical structure of the ensemble for intervals of that duration.

20. Entropy of a Continuous Distribution

The entropy of a discrete set of probabilities $p_1, \cdot \cdot \cdot, p_n$ has been defined as:

$$H = -\sum p_i \log p_i.$$

In an analogous manner we define the entropy of a continuous distribution with the density distribution function $p(x)$ by:

$$H = -\int_{-\infty}^{\infty} p(x) \log p(x) \, dx.$$

With an n dimensional distribution $p(x_1, \cdot \cdot \cdot, x_n)$ we have

$$H = -\int \cdot \cdot \cdot \int p(x_1 \cdot \cdot \cdot x_n) \log p(x_1, \cdot \cdot \cdot, x_n) \, dx_1 \cdot \cdot \cdot dx_n.$$

If we have two arguments x and y (which may themselves be multidimensional) the joint and conditional entropies of $p(x, y)$ are given by

$$H(x, y) = -\iint p(x, y) \log p(x, y) \, dx \, dy$$

and

$$H_x(y) = -\iint p(x, y) \log \frac{p(x, y)}{p(x)} \, dx \, dy$$

$$H_y(x) = -\iint p(x, y) \log \frac{p(x, y)}{p(y)} \, dx \, dy$$

where

$$p(x) = \int p(x, y) \, dy$$

$$p(y) = \int p(x, y) \, dx.$$

The entropies of continuous distributions have most (but not all) of the properties of the discrete case. In particular we have the following:

1. If x is limited to a certain volume v in its space, then $H(x)$ is a maximum and equal to $\log v$ when $p(x)$ is constant $\left(\frac{1}{v}\right)$ in the volume.

2. With any two variables x, y we have

$$H(x, y) \leq H(x) + H(y)$$

with equality if (and only if) x and y are independent, i.e., $p(x, y) = p(x) \, p(y)$ (apart possibly from a set of points of probability zero).

3. Consider a generalized averaging operation of the following type:

$$p'(y) = \int a(x, y) \, p(x) \, dx$$

with

$$\int a(x, y) \, dx = \int a(x, y) \, dy = 1, \qquad a(x, y) \geq 0.$$

Then the entropy of the averaged distribution $p'(y)$ is equal to or greater than that of the original distribution $p(x)$.

4. We have

$$H(x, y) = H(x) + H_x(y) = H(y) + H_y(x)$$

and

$$H_x(y) \leq H(y).$$

5. Let $p(x)$ be a one-dimensional distribution. The form of $p(x)$ giving a maximum entropy subject to the condition that the standard deviation of x be fixed at σ is Gaussian. To show this we must maximize

$$H(x) = -\int p(x) \log p(x) \, dx$$

with

$$\sigma^2 = \int p(x) x^2 \, dx \qquad \text{and} \qquad 1 = \int p(x) \, dx$$

as constraints. This requires, by the calculus of variations, maximizing

$$\int [-p(x) \log p(x) + \lambda p(x) x^2 + \mu p(x)] \, dx.$$

The condition for this is

$$-1 - \log p(x) + \lambda x^2 + \mu = 0$$

and consequently (adjusting the constants to satisfy the constraints)

$$p(x) = \frac{1}{\sqrt{2\pi}\,\sigma}\, e^{-(x^2/2\sigma^2)}.$$

Similarly in n dimensions, suppose the second order moments of $p(x_1, \cdots, x_n)$ are fixed at A_{ij}:

$$A_{ij} = \int \cdots \int x_i x_j p(x_i, \cdots, x_n)\, dx_1 \cdots dx_n.$$

Then the maximum entropy occurs (by a similar calculation) when $p(x_1, \cdots, x_n)$ is the n dimensional Gaussian distribution with the second order moments A_{ij}.

6. The entropy of a one-dimensional Gaussian distribution whose standard deviation is σ is given by

$$H(x) = \log \sqrt{2\pi e}\, \sigma.$$

This is calculated as follows:

$$p(x) = \frac{1}{\sqrt{2\pi}\,\sigma}\, e^{-(x^2/2\sigma^2)}$$

$$-\log p(x) = \log \sqrt{2\pi}\,\sigma + \frac{x^2}{2\sigma^2}$$

$$H(x) = -\int p(x) \log p(x)\, dx$$

$$= \int p(x) \log \sqrt{2\pi}\,\sigma\, dx + \int p(x)\, \frac{x^2}{2\sigma^2}\, dx$$

$$= \log \sqrt{2\pi}\,\sigma + \frac{\sigma^2}{2\sigma^2}$$

$$= \log \sqrt{2\pi}\,\sigma + \log \sqrt{e}$$

$$= \log \sqrt{2\pi e}\, \sigma.$$

Similarly the n dimensional Gaussian distribution with associated quadratic form a_{ij} is given by

$$p(x_1, \cdots, x_n) = \frac{|a_{ij}|^{\frac{1}{2}}}{(2\pi)^{n/2}} \exp\left(-\tfrac{1}{2}\Sigma a_{ij}x_i x_j\right)$$

and the entropy can be calculated as

$$H = \log \, (2\pi e)^{n/2} \, |\, a_{ij} \,|^{-\frac{1}{2}}$$

where $|\, a_{ij} \,|$ is the determinant whose elements are a_{ij}.

7. If x is limited to a half line ($p(x) = 0$ for $x \leq 0$) and the first moment of x is fixed at a:

$$a = \int_0^\infty p(x)x \, dx,$$

then the maximum entropy occurs when

$$p(x) = \frac{1}{a} \, e^{-(x/a)}$$

and is equal to log *ea*.

8. There is one important difference between the continuous and discrete entropies. In the discrete case the entropy measures in an *absolute* way the randomness of the chance variable. In the continuous case the measurement is *relative to the coordinate system*. If we change coordinates the entropy will in general change. In fact if we change to coordinates $y_1 \cdots y_n$ the new entropy is given by

$$H(y) = \int \cdots \int p(x_1 \cdots x_n) \, J\left(\frac{x}{y}\right)$$

$$\log \, p(x_1 \cdots x_n) \, J\left(\frac{x}{y}\right) dy_1 \cdots dy_n$$

where $J\left(\dfrac{x}{y}\right)$ is the Jacobian of the coordinate transformation. On expanding the logarithm and changing variables to $x_1 \cdots x_n$, we obtain:

$$H(y) = H(x) - \int \cdots \int p(x_1, \cdots, x_n) \log \, J\left(\frac{x}{y}\right) dx_1 \cdots dx_n.$$

Thus the new entropy is the old entropy less the expected logarithm of the Jacobian. In the continuous case the entropy can be considered a measure of randomness *relative to an assumed standard*, namely the coordinate system chosen with each small volume element $dx_1 \cdots dx_n$ given equal weight. When we change the coordinate system the entropy in the new system measures the randomness when equal volume elements $dy_1 \cdots dy_n$ in the new system are given equal weight.

In spite of this dependence on the coordinate system the

entropy concept is as important in the continuous case as the discrete case. This is due to the fact that the derived concepts of information rate and channel capacity depend on the *difference* of two entropies and this difference *does not* depend on the coordinate frame, each of the two terms being changed by the same amount.

The entropy of a continuous distribution can be negative. The scale of measurements sets an arbitrary zero corresponding to a uniform distribution over a unit volume. A distribution which is more confined than this has less entropy and will be negative. The rates and capacities will, however, always be non-negative.

9. A particular case of changing coordinates is the linear transformation

$$y_j = \sum_i a_{ij} x_i.$$

In this case the Jacobian is simply the determinant $\mid a_{ij} \mid^{-1}$ and

$$H(y) = H(x) + \log \mid a_{ii} \mid.$$

In the case of a rotation of coordinates (or any measure preserving transformation) $J = 1$ and $H(y) = H(x)$.

21. Entropy of an Ensemble of Functions

Consider an ergodic ensemble of functions limited to a certain band of width W cycles per second. Let

$$p(x_1, \cdots, x_n)$$

be the density distribution function for amplitudes $x_1 \cdots x_n$ at n successive sample points. We define the entropy of the ensemble per degree of freedom by

$$H' = - \operatorname*{Lim}_{n \to \infty} \frac{1}{n} \int \cdots \int p(x_1, \cdots, x_n) \\ \log p(x_1, \cdots, x_n) \, dx_1 \cdots dx_n.$$

We may also define an entropy H per second by dividing, not by n, but by the time T in seconds for n samples. Since $n = 2TW$, $H = 2WH'$.

With white thermal noise p is Gaussian and we have

$$H' = \log \sqrt{2\pi eN},$$
$$H = W \log 2\pi eN.$$

For a given average power N, white noise has the maximum possible entropy. This follows from the maximizing properties of the Gaussian distribution noted above.

The entropy for a continuous stochastic process has many properties analogous to that for discrete processes. In the discrete case the entropy was related to the logarithm of the *probability* of long sequences, and to the *number* of reasonably probable sequences of long length. In the continuous case it is related in a similar fashion to the logarithm of the *probability density* for a long series of samples, and the *volume* of reasonably high probability in the function space.

More precisely, if we assume $p(x_1, \cdots, x_n)$ continuous in all the x_i for all n, then for sufficiently large n

$$\left| \frac{\log p}{n} - H' \right| < \epsilon$$

for all choices of $(x_1, \cdots, x_n)$ apart from a set whose total probability is less than δ, with δ and ϵ arbitrarily small. This follows from the ergodic property if we divide the space into a large number of small cells.

The relation of H to volume can be stated as follows: Under the same assumptions consider the n dimensional space corresponding to $p(x_1, \cdots, x_n)$. Let $V_n(q)$ be the smallest volume in this space which includes in its interior a total probability q. Then

$$\lim_{n \to \infty} \frac{\log V_n(q)}{n} = H'$$

provided q does not equal 0 or 1.

These results show that for large n there is a rather well-defined volume (at least in the logarithmic sense) of high probability, and that within this volume the probability density is relatively uniform (again in the logarithmic sense).

In the white noise case the distribution function is given by

$$p(x_1, \cdots, x_n) = \frac{1}{(2\pi N)^{n/2}} \exp -\frac{1}{2N} \Sigma x_i^2.$$

Since this depends only on Σx_i^2 the surfaces of equal probability density are spheres and the entire distribution has spherical symmetry. The region of high probability is a sphere of radius $\sqrt{nN}$. As $n \to \infty$ the probability of being outside a sphere of radius $\sqrt{n(N + \epsilon)}$ approaches zero however small ϵ and $\frac{1}{n}$ times the logarithm of the volume of the sphere approaches $\log \sqrt{2\pi eN}$.

In the continuous case it is convenient to work not with the entropy H of an ensemble but with a derived quantity which we will call the *entropy power*. This is defined as the power in a white noise limited to the same band as the original ensemble and having the same entropy. In other words if H' is the entropy of an ensemble its entropy power is

$$N_1 = \frac{1}{2\pi e} \exp 2H'.$$

In the geometrical picture this amounts to measuring the high probability volume by the squared radius of a sphere having the same volume. Since white noise has the maximum entropy for a given power, the entropy power of any noise is less than or equal to its actual power.

22. Entropy Loss in Linear Filters

Theorem 14: If an ensemble having an entropy H_1 per degree of freedom in band W is passed through a filter with characteristic $Y(f)$ the output ensemble has an entropy

$$H_2 = H_1 + \frac{1}{W} \int_W \log |Y(f)|^2 \, df.$$

The operation of the filter is essentially a linear transformation of coordinates. If we think of the different frequency components as the original coordinate system, the new frequency components are merely the old ones multiplied by factors. The coordinate transformation matrix is thus essentially diagonalized in terms of these coordinates. The Jacobian of the transformation is (for n sine and n cosine components)

$$J = \prod_{i=1}^{n} |Y(f_i)|^2 = \exp \Sigma \log |Y(f_i)|^2$$

where the f_i are equally spaced through the band W. This becomes in the limit

$$\exp \frac{1}{W} \int_W \log \mid Y(f) \mid^2 df.$$

Since J is constant its average value is the same quantity and applying the theorem on the change of entropy with a change of

TABLE I

GAIN	ENTROPY POWER FACTOR	ENTROPY POWER GAIN IN DECIBELS	IMPULSE RESPONSE TIMES π
$1-\omega$ ---->	$\dfrac{1}{e^2}$	-8.68	$\dfrac{\text{SIN}^2\,(t/2)}{t^2/2}$
$1-\omega^2$ ---->	$\left(\dfrac{2}{e}\right)^4$	-5.33	$2\left[\dfrac{\text{SIN}\,t}{t^3} - \dfrac{\cos t}{t^2}\right]$
$1-\omega^3$ ---->	0.411	-3.87	$6\left[\dfrac{\cos t - 1}{t^4} - \dfrac{\cos t}{2t^2} + \dfrac{\text{SIN}\,t}{t^3}\right]$
$\sqrt{1-\omega^2}$ ---->	$\left(\dfrac{2}{e}\right)^2$	-2.67	$\dfrac{\pi}{2}\,\dfrac{J_1\,(t)}{t}$
	$\dfrac{1}{e^{2\alpha}}$	$-8.68\,\alpha$	$\dfrac{1}{\alpha t^2}\left[\cos\,(1-\alpha)t-\cos t\right]$

coordinates, the result follows. We may also phrase it in terms of the entropy power. Thus if the entropy power of the first ensemble is N_1 that of the second is

$$N_1 \exp \frac{1}{W} \int_W \log \mid Y(f) \mid^2 df.$$

The final entropy power is the initial entropy power multiplied by the geometric mean gain of the filter. If the gain is measured in *db*, then the output entropy power will be increased by the arithmetic mean *db* gain over W.

In Table I the entropy power loss has been calculated (and also expressed in *db*) for a number of ideal gain characteristics. The impulsive responses of these filters are also given for $W = 1/2\pi$, with phase assumed to be 0.

The entropy loss for many other cases can be obtained from these results. For example the entropy power factor $\frac{1}{e^2}$ for the first case also applies to any gain characteristic obtained from $1 - \omega$ by a measure preserving transformation of the ω axis. In particular a linearly increasing gain $G(\omega) = \omega$, or a "saw tooth" characteristic between 0 and 1 have the same entropy loss. The reciprocal gain has the reciprocal factor. Thus $\frac{1}{\omega}$ has the factor e^2. Raising the gain to any power raises the factor to this power.

23. Entropy of the Sum of Two Ensembles

If we have two ensembles of functions $f_\alpha(t)$ and $g_\beta(t)$ we can form a new ensemble by "addition." Suppose the first ensemble has the probability density function $p(x_1, \cdots, x_n)$ and the second $q(x_1, \cdots, x_n)$. Then the density function for the sum is given by the convolution:

$$r(x_1, \cdots, x_n) = \int \cdots \int p(y_1, \cdots, y_n)$$
$$\cdot q(x_1 - y_1, \cdots, x_n - y_n) \, dy_1 \, dy_2 \cdots dy_n.$$

Physically this corresponds to adding the noises or signals represented by the original ensembles of functions.

The following result is derived in Appendix 6.

Theorem 15: Let the average power of two ensembles be N_1

and N_2 and let their entropy powers be $\overline{N}_1$ and $\overline{N}_2$. Then the entropy power of the sum, $\overline{N}_3$, is bounded by

$$\overline{N}_1 + \overline{N}_2 \leq \overline{N}_3 \leq N_1 + N_2.$$

White Gaussian noise has the peculiar property that it can absorb any other noise or signal ensemble which may be added to it with a resultant entropy power approximately equal to the sum of the white noise power and the signal power (measured from the average signal value, which is normally zero), provided the signal power is small, in a certain sense, compared to the noise.

Consider the function space associated with these ensembles having n dimensions. The white noise corresponds to the spherical Gaussian distribution in this space. The signal ensemble corresponds to another probability distribution, not necessarily Gaussian or spherical. Let the second moments of this distribution about its center of gravity be a_{ij}. That is, if $p(x_1, \cdots , x_n)$ is the density distribution function

$$a_{ij} = \int \cdots \int p(x_i - \alpha_i)\ (x_j - \alpha_j)\ dx_1 \cdots dx_n$$

where the α_i are the coordinates of the center of gravity. Now a_{ij} is a positive definite quadratic form, and we can rotate our coordinate system to align it with the principal directions of this form. a_{ij} is then reduced to diagonal form b_{ii}. We require that each b_{ii} be small compared to N, the squared radius of the spherical distribution.

In this case the convolution of the noise and signal produce approximately a Gaussian distribution whose corresponding quadratic form is

$$N + b_{ii}.$$

The entropy power of this distribution is

$$[\Pi(N + b_{ii})]^{1/n}$$

or approximately

$$= [(N)^n + \Sigma b_{ii}(N)^{n-1}]^{1/n}$$

$$= N + \frac{1}{n}\ \Sigma b_{ii}.$$

The last term is the signal power, while the first is the noise power.

IV

The Continuous Channel

24. The Capacity of a Continuous Channel

In a continuous channel the input or transmitted signals will be continuous functions of time $f(t)$ belonging to a certain set, and the output or received signals will be perturbed versions of these. We will consider only the case where both transmitted and received signals are limited to a certain band W. They can then be specified, for a time T, by $2TW$ numbers, and their statistical structure by finite dimensional distribution functions. Thus the statistics of the transmitted signal will be determined by

$$P(x_1, \cdots, x_n) = P(x)$$

and those of the noise by the conditional probability distribution

$$P_{x_1, \cdots, x_n}(y_1, \cdots, y_n) = P_x(y).$$

The rate of transmission of information for a continuous channel is defined in a way analogous to that for a discrete channel, namely

$$R = H(x) - H_y(x)$$

where $H(x)$ is the entropy of the input and $H_y(x)$ the equivocation. The channel capacity C is defined as the maximum of R when we vary the input over all possible ensembles. This means that in a finite dimensional approximation we must vary $P(x) \doteq P(x_1, \cdots, x_n)$ and maximize

$$- \int P(x) \log P(x) \, dx + \int\int P(x, y) \log \frac{P(x, y)}{P(y)} \, dx \, dy.$$

This can be written

$$\iint P(x, y) \log \frac{P(x, y)}{P(x)P(y)} \, dx \, dy$$

using the fact that $\iint P(x, y) \log P(x) \, dx \, dy = \int P(x) \log P(x) \, dx$.
The channel capacity is thus expressed as follows:

$$C = \underset{T \to \infty}{\text{Lim}} \ \underset{P(x)}{\text{Max}} \ \frac{1}{T} \iint P(x, y) \log \frac{P(x, y)}{P(x)P(y)} \, dx \, dy.$$

It is obvious in this form that R and C are independent of the coordinate system since the numerator and denominator in log $\frac{P(x, y)}{P(x)P(y)}$ will be multiplied by the same factors when x and y are transformed in any one-to-one way. This integral expression for C is more general than $H(x) - H_y(x)$. Properly interpreted (see Appendix 7) it will always exist while $H(x) - H_y(x)$ may assume an indeterminate form $\infty - \infty$ in some cases. This occurs, for example, if x is limited to a surface of fewer dimensions than n in its n dimensional approximation.

If the logarithmic base used in computing $H(x)$ and $H_y(x)$ is two then C is the maximum number of binary digits that can be sent per second over the channel with arbitrarily small equivocation, just as in the discrete case. This can be seen physically by dividing the space of signals into a large number of small cells, sufficiently small so that the probability density $P_x(y)$ of signal x being perturbed to point y is substantially constant over a cell (either of x or y). If the cells are considered as distinct points the situation is essentially the same as a discrete channel and the proofs used there will apply. But it is clear physically that this quantizing of the volume into individual points cannot in any practical situation alter the final answer significantly, provided the regions are sufficiently small. Thus the capacity will be the limit of the capacities for the discrete subdivisions and this is just the continuous capacity defined above.

On the mathematical side it can be shown first (see Appendix 7) that if u is the message, x is the signal, y is the received signal (perturbed by noise) and v the recovered message then

$$H(x) - H_y(x) \geq H(u) - H_v(u)$$

regardless of what operations are performed on u to obtain x or on y to obtain v. Thus no matter how we encode the binary digits to obtain the signal, or how we decode the received signal to recover the message, the discrete rate for the binary digits does not exceed the channel capacity we have defined. On the other hand, it is possible under very general conditions to find a coding system for transmitting binary digits at the rate C with as small an equivocation or frequency of errors as desired. This is true, for example, if, when we take a finite dimensional approximating space for the signal functions, $P(x, y)$ is continuous in both x and y except at a set of points of probability zero.

An important special case occurs when the noise is added to the signal and is independent of it (in the probability sense). Then $P_x(y)$ is a function only of the (vector) difference $n = (y - x)$,

$$P_x(y) = Q(y - x)$$

and we can assign a definite entropy to the noise (independent of the statistics of the signal), namely the entropy of the distribution $Q(n)$. This entropy will be denoted by $H(n)$.

Theorem 16: If the signal and noise are independent and the received signal is the sum of the transmitted signal and the noise then the rate of transmission is

$$R = H(y) - H(n),$$

i.e., the entropy of the received signal less the entropy of the noise. The channel capacity is

$$C = \max_{P(x)} H(y) - H(n).$$

We have, since $y = x + n$:

$$H(x, y) = H(x, n).$$

Expanding the left side and using the fact that x and n are independent

$$H(y) + H_y(x) = H(x) + H(n).$$

Hence

$$R = H(x) - H_y(x) = H(y) - H(n).$$

Since $H(n)$ is independent of $P(x)$, maximizing R requires maximizing $H(y)$, the entropy of the received signal. If there are certain constraints on the ensemble of transmitted signals, the entropy of the received signal must be maximized subject to these constraints.

25. Channel Capacity with an Average Power Limitation

A simple application of Theorem 16 occurs when the noise is a white thermal noise and the transmitted signals are limited to a certain average power P. Then the received signals have an average power $P + N$ where N is the average noise power. The maximum entropy for the received signals occurs when they also form a white noise ensemble since this is the greatest possible entropy for a power $P + N$ and can be obtained by a suitable choice of the ensemble of transmitted signals, namely if they form a white noise ensemble of power P. The entropy (per second) of the received ensemble is then

$$H(y) = W \log 2\pi e (P + N),$$

and the noise entropy is

$$H(n) = W \log 2\pi e N.$$

The channel capacity is

$$C = H(y) - H(n) = W \log \frac{P + N}{N}.$$

Summarizing we have the following:

Theorem 17: The capacity of a channel of band W perturbed by white thermal noise of power N when the average transmitter power is limited to P is given by

$$C = W \log \frac{P + N}{N}.$$

This means that by sufficiently involved encoding systems we can transmit binary digits at the rate $W \log_2 \dfrac{P + N}{N}$ bits per second, with arbitrarily small frequency of errors. It is not possible to transmit at a higher rate by any encoding system without a definite positive frequency of errors.

To approximate this limiting rate of transmission the transmitted signals must approximate, in statistical properties, a white noise.[6] A system which approaches the ideal rate may be described as follows: Let $M = 2^s$ samples of white noise be constructed each of duration T. These are assigned binary numbers from 0 to $(M - 1)$. At the transmitter the message sequences are broken up into groups of s and for each group the corresponding noise sample is transmitted as the signal. At the receiver the M samples are known and the actual received signal (perturbed by noise) is compared with each of them. The sample which has the least R.M.S. discrepancy from the received signal is chosen as the transmitted signal and the corresponding binary number reconstructed. This process amounts to choosing the most probable (*a posteriori*) signal. The number M of noise samples used will depend on the tolerable frequency ϵ of errors, but for almost all selections of samples we have

$$\underset{\epsilon \to 0}{\text{Lim}} \; \underset{T \to \infty}{\text{Lim}} \; \frac{\log M(\epsilon, T)}{T} = W \log \frac{P + N}{N},$$

so that no matter how small ϵ is chosen, we can, by taking T sufficiently large, transmit as near as we wish to $TW \log \dfrac{P + N}{N}$ binary digits in the time T.

Formulas similar to $C = W \log \dfrac{P + N}{N}$ for the white noise case have been developed independently by several other writers, although with somewhat different interpretations. We may mention the work of N. Wiener,[7] W. G. Tuller,[8] and H. Sullivan in this connection.

In the case of an arbitrary perturbing noise (not necessarily white thermal noise) it does not appear that the maximizing problem involved in determining the channel capacity C can be solved explicitly. However, upper and lower bounds can be set for C in terms of the average noise power N and the noise entropy power N_1. These bounds are sufficiently close together in most

---

[6] This and other properties of the white noise case are discussed from the geometrical point of view in "Communication in the Presence of Noise," *loc. cit.*

[7] *Cybernetics, loc. cit.*

[8] "Theoretical Limitations on the Rate of Transmission of Information," *Proceedings of the Institute of Radio Engineers,* v. 37, No. 5, May, 1949, pp. 468-78.

practical cases to furnish a satisfactory solution to the problem.

Theorem 18: The capacity of a channel of band W perturbed by an arbitrary noise is bounded by the inequalities

$$W \log \frac{P + N_1}{N_1} \leq C \leq W \log \frac{P + N}{N_1}$$

where

P = *average transmitter power*
N = *average noise power*
N_1 = *entropy power of the noise.*

Here again the average power of the perturbed signals will be $P + N$. The maximum entropy for this power would occur if the received signal were white noise and would be $W \log 2\pi e (P + N)$. It may not be possible to achieve this; i.e., there may not be any ensemble of transmitted signals which, added to the perturbing noise, produce a white thermal noise at the receiver, but at least this sets an upper bound to $H(y)$. We have, therefore

$$C = \text{Max } H(y) - H(n)$$
$$\leq W \log 2\pi e (P + N) - W \log 2\pi e N_1.$$

This is the upper limit given in the theorem. The lower limit can be obtained by considering the rate if we make the transmitted signal a white noise, of power P. In this case the entropy power of the received signal must be at least as great as that of a white noise of power $P + N_1$ since we have shown in Theorem 15 that the entropy power of the sum of two ensembles is greater than or equal to the sum of the individual entropy powers. Hence

$$\text{Max } H(y) \geq W \log 2\pi e (P + N_1)$$

and

$$C \geq W \log 2\pi e (P + N_1) - W \log 2\pi e N_1$$
$$= W \log \frac{P + N_1}{N_1}.$$

As P increases, the upper and lower bounds in Theorem 18 approach each other, so we have as an asymptotic rate

$$W \log \frac{P + N}{N_1}.$$

If the noise is itself white, $N = N_1$ and the result reduces to the formula proved previously:

$$C = W \log \left(1 + \frac{P}{N}\right).$$

If the noise is Gaussian but with a spectrum which is not necessarily flat, N_1 is the geometric mean of the noise power over the various frequencies in the band W. Thus

$$N_1 = \exp \frac{1}{W} \int_W \log N(f) \, df$$

where $N(f)$ is the noise power at frequency f.

Theorem 19: If we set the capacity for a given transmitter power P equal to

$$C = W \log \frac{P + N - \eta}{N_1}$$

then η is monotonic decreasing as P increases and approaches 0 as a limit.

Suppose that for a given power P_1 the channel capacity is

$$W \log \frac{P_1 + N - \eta_1}{N_1}.$$

This means that the best signal distribution, say $p(x)$, when added to the noise distribution $q(x)$, gives a received distribution $r(y)$ whose entropy power is $(P_1 + N - \eta_1)$. Let us increase the power to $P_1 + \Delta P$ by adding a white noise of power ΔP to the signal. The entropy of the received signal is now at least

$$H(y) = W \log 2\pi e (P_1 + N - \eta_1 + \Delta P)$$

by application of the theorem on the minimum entropy power of a sum. Hence, since we can attain the H indicated, the entropy of the maximizing distribution must be at least as great and η must be monotonic decreasing. To show that $\eta \to 0$ as $P \to \infty$ consider a signal which is a white noise with a large P. Whatever the perturbing noise, the received signal will be approximately a white noise, if P is sufficiently large, in the sense of having an entropy power approaching $P + N$.

26. The Channel Capacity with a Peak Power Limitation

In some applications the transmitter is limited not by the average power output but by the peak instantaneous power. The problem of calculating the channel capacity is then that of maximizing (by variation of the ensemble of transmitted symbols)

$$H(y) - H(n)$$

subject to the constraint that all the functions $f(t)$ in the ensemble be less than or equal to $\sqrt{S}$, say, for all t. A constraint of this type does not work out as well mathematically as the average power limitation. The most we have obtained for this case is a lower bound valid for all $\frac{S}{N}$, an "asymptotic" upper bound $\left(\text{valid for large } \frac{S}{N}\right)$ and an asymptotic value of C for $\frac{S}{N}$ small.

Theorem 20: The channel capacity C for a band W perturbed by white thermal noise of power N is bounded by

$$C \geq W \log \frac{2}{\pi e^3} \frac{S}{N},$$

where S is the peak allowed transmitter power. For sufficiently large $\frac{S}{N}$

$$C \leq W \log \frac{\frac{2}{\pi e} S + N}{N} (1 + \epsilon)$$

where ϵ is arbitrarily small. As $\frac{S}{N} \to 0$ (and provided the band W starts at 0)

$$C / W \log \left(1 + \frac{S}{N}\right) \to 1.$$

We wish to maximize the entropy of the received signal. If $\frac{S}{N}$ is large this will occur very nearly when we maximize the entropy of the transmitted ensemble.

The asymptotic upper bound is obtained by relaxing the conditions on the ensemble. Let us suppose that the power is limited to S not at every instant of time, but only at the sample points. The maximum entropy of the transmitted ensemble under these weakened conditions is certainly greater than or equal to that under the original conditions. This altered problem can be solved easily. The maximum entropy occurs if the different samples are independent and have a distribution function which is constant from $-\sqrt{S}$ to $+\sqrt{S}$. The entropy can be calculated as

$$W \log 4S.$$

The received signal will then have an entropy less than

$$W \log (4S + 2\pi eN)(1 + \epsilon)$$

with $\epsilon \to 0$ as $\dfrac{S}{N} \to \infty$ and the channel capacity is obtained by subtracting the entropy of the white noise, $W \log 2\pi eN$:

$$W \log (4S + 2\pi eN) (1 + \epsilon) - W \log (2\pi eN)$$

$$= W \log \frac{\dfrac{2}{\pi e} S + N}{N} (1 + \epsilon).$$

This is the desired upper bound to the channel capacity.

To obtain a lower bound consider the same ensemble of functions. Let these functions be passed through an ideal filter with a triangular transfer characteristic. The gain is to be unity at frequency 0 and decline linearly down to gain 0 at frequency W. We first show that the output functions of the filter have a peak power limitation S at all times (not just the sample points). First we note that a pulse $\dfrac{\sin 2\pi Wt}{2\pi Wt}$ going into the filter produces

$$\frac{1}{2} \frac{\sin^2 \pi Wt}{(\pi Wt)^2}$$

in the output. This function is never negative. The input function (in the general case) can be thought of as the sum of a series of shifted functions

$$a \frac{\sin 2\pi Wt}{2\pi Wt}$$

where a, the amplitude of the sample, is not greater than $\sqrt{S}$. Hence the output is the sum of shifted functions of the nonnegative form above with the same coefficients. These functions being non-negative, the greatest positive value for any t is obtained when all the coefficients a have their maximum positive values, i.e., $\sqrt{S}$. In this case the input function was a constant of amplitude $\sqrt{S}$ and since the filter has unit gain for D.C., the output is the same. Hence the output ensemble has a peak power S.

The entropy of the output ensemble can be calculated from that of the input ensemble by using the theorem dealing with such a situation. The output entropy is equal to the input entropy plus the geometrical mean gain of the filter:

$$\int_0^W \log G^2 \, df = \int_0^W \log \left(\frac{W - f}{W} \right)^2 df = -2W.$$

Hence the output entropy is

$$W \log 4S - 2W = W \log \frac{4S}{e^2}$$

and the channel capacity is greater than

$$W \log \frac{2}{\pi e^3} \frac{S}{N}.$$

We now wish to show that, for small $\dfrac{S}{N}$ (peak signal power over average white noise power), the channel capacity is approximately

$$C = W \log \left(1 + \frac{S}{N} \right).$$

More precisely $C / W \log \left(1 + \dfrac{S}{N} \right) \to 1$ as $\dfrac{S}{N} \to 0$. Since the average signal power P is less than or equal to the peak S, it follows that for all $\dfrac{S}{N}$

$$C \le W \log \left(1 + \frac{P}{N} \right) \le W \log \left(1 + \frac{S}{N} \right).$$

Therefore, if we can find an ensemble of functions such that they correspond to a rate nearly $W \log \left(1 + \dfrac{S}{N} \right)$ and are limited to band W and peak S the result will be proved. Consider the ensemble of functions of the following type. A series of t samples have the same value, either $+\sqrt{S}$ or $-\sqrt{S}$, then the next t samples have the same value, etc. The value for a series is chosen at random, probability $\frac{1}{2}$ for $+\sqrt{S}$ and $\frac{1}{2}$ for $-\sqrt{S}$. If this ensemble be passed through a filter with triangular gain characteristic (unit gain at D.C.), the output is peak limited to $\pm S$. Furthermore the average power is nearly S and can be made to approach this by taking t sufficiently large. The entropy of the sum of this and the thermal noise can be found by applying the theorem on the sum of a noise and a small signal. This theorem will apply if

$$\sqrt{t}\ \frac{S}{N}$$

is sufficiently small. This can be ensured by taking $\frac{S}{N}$ small enough (after t is chosen). The entropy power will be $S+N$ to as close an approximation as desired, and hence the rate of transmission as near as we wish to

$$W \log \left(\frac{S+N}{N} \right).$$

V

The Rate for a Continuous Source

27. Fidelity Evaluation Functions

In the case of a discrete source of information we were able to determine a definite rate of generating information, namely the entropy of the underlying stochastic process. With a continuous source the situation is considerably more involved. In the first place a continuously variable quantity can assume an infinite number of values and requires, therefore, an infinite number of binary digits for exact specification. This means that to transmit the output of a continuous source with *exact recovery* at the receiving point requires, in general, a channel of infinite capacity (in bits per second). Since, ordinarily, channels have a certain amount of noise, and therefore a finite capacity, exact transmission is impossible.

This, however, evades the real issue. Practically, we are not interested in exact transmission when we have a continuous source, but only in transmission to within a certain tolerance. The question is, can we assign a definite rate to a continuous source when we require only a certain fidelity of recovery, measured in a suitable way. Of course, as the fidelity requirements are increased the rate will increase. It will be shown that we can, in very general cases, define such a rate, having the property that it is possible, by properly encoding the information, to transmit it over a channel whose capacity is equal to the rate in question, and satisfy the fidelity requirements. A channel of smaller capacity is insufficient.

It is first necessary to give a general mathematical formulation of the idea of fidelity of transmission. Consider the set of messages of a long duration, say T seconds. The source is described by giving the probability density, $P(x)$, in the associated space, that the source will select the message in question. A given communication system is described (from the external point of view) by giving the conditional probability $P_x(y)$ that if message x is produced by the source the recovered message at the receiving point will be y. The system as a whole (including source and transmission system) is described by the probability function $P(x, y)$ of having message x and final output y. If this function is known, the complete characteristics of the system from the point of view of fidelity are known. Any evaluation of fidelity must correspond mathematically to an operation applied to $P(x, y)$. This operation must at least have the properties of a simple ordering of systems; i.e., it must be possible to say of two systems represented by $P_1(x, y)$ and $P_2(x, y)$ that, according to our fidelity criterion, either (1) the first has higher fidelity, (2) the second has higher fidelity, or (3) they have equal fidelity. This means that a criterion of fidelity can be represented by a numerically valued *evaluation function*:

$$v(P(x, y))$$

whose argument ranges over possible probability functions $P(x, y)$. The function $v(P(x, y))$ orders communication systems according to fidelity, and for convenience we take lower values of v to correspond to "higher fidelity."

We will now show that under very general and reasonable assumptions the function $v(P(x, y))$ can be written in a seemingly much more specialized form, namely as an average of a function $\rho(x, y)$ over the set of possible values of x and y:

$$v(P(x, y)) = \iint P(x, y)\ \rho(x, y)\ dx\ dy.$$

To obtain this we need only assume (1) that the source and system are ergodic so that a very long sample will be, with probability nearly 1, typical of the ensemble, and (2) that the evaluation is "reasonable" in the sense that it is possible, by observing a typical input and output x_1 and y_1, to form a tentative evaluation on the basis of these samples; and if these samples are increased in duration the tentative evaluation will, with probability 1, approach the exact evaluation based on a full knowledge of $P(x, y)$. Let the tentative evaluation be $\rho(x, y)$. Then the function $\rho(x, y)$ approaches (as $T \to \infty$) a constant for almost all (x, y) which are in the high probability region corresponding to the system:

$$\rho(x, y) \to v(P(x, y))$$

and we may also write

$$\rho(x, y) \to \iint P(x, y) \; \rho(x, y) \; dx \; dy$$

since

$$\iint P(x, y) \; dx \; dy = 1.$$

This establishes the desired result.

The function $\rho(x, y)$ has the general nature of a "distance" between x and y.[9] It measures how undesirable it is (according to our fidelity criterion) to receive y when x is transmitted. The general result given above can be restated as follows: Any reasonable evaluation can be represented as an average of a distance function over the set of messages and recovered messages x and y weighted according to the probability $P(x, y)$ of getting the pair in question, provided the duration T of the messages be taken sufficiently large.

The following are simple examples of evaluation functions:

1. R.M.S. criterion.

$$v = \overline{(x(t) - y(t))^2}.$$

In this very commonly used measure of fidelity the distance function $\rho(x, y)$ is (apart from a constant factor) the square of the ordinary Euclidean distance between the points x and y in the associated function space.

$$\rho(x, y) = \frac{1}{T} \int_0^T [x(t) - y(t)]^2 \; dt.$$

2. Frequency weighted R.M.S. criterion. More generally one can apply different weights to the different frequency components

---

[9] It is not a "metric" in the strict sense, however, since in general it does not satisfy either $\rho(x, y) = \rho(y, x)$ or $\rho(x, y) + \rho(y, z) \geq \rho(x, z)$.

before using an R.M.S. measure of fidelity. This is equivalent to passing the difference $x(t) - y(t)$ through a shaping filter and then determining the average power in the output. Thus let

$$e(t) = x(t) - y(t)$$

and

$$f(t) = \int_{-\infty}^{\infty} e(\tau)k(t - \tau) \, d\tau$$

then

$$\rho(x, y) = \frac{1}{T} \int_0^T f(t)^2 \, dt.$$

3. Absolute error criterion.

$$\rho(x, y) = \frac{1}{T} \int_0^T | \, x(t) - y(t) \, | \, dt.$$

4. The structure of the ear and brain determine implicitly a number of evaluations, appropriate in the case of speech or music transmission. There is, for example, an "intelligibility" criterion in which $\rho(x, y)$ is equal to the relative frequency of incorrectly interpreted words when message $x(t)$ is received as $y(t)$. Although we cannot give an explicit representation of $\rho(x, y)$ in these cases it could, in principle, be determined by sufficient experimentation. Some of its properties follow from well-known experimental results in hearing, e.g., the ear is relatively insensitive to phase and the sensitivity to amplitude and frequency is roughly logarithmic.

5. The discrete case can be considered as a specialization in which we have tacitly assumed an evaluation based on the frequency of errors. The function $\rho(x, y)$ is then defined as the number of symbols in the sequence y differing from the corresponding symbols in x divided by the total number of symbols in x.

28. The Rate for a Source Relative to a Fidelity Evaluation

We are now in a position to define a rate of generating information for a continuous source. We are given $P(x)$ for the source and an evaluation v determined by a distance function $\rho(x, y)$ which will be assumed continuous in both x and y. With a particular system $P(x, y)$ the quality is measured by

$$v = \iint \rho(x, y) \, P(x, y) \, dx \, dy.$$

Furthermore the rate of flow of binary digits corresponding to $P(x, y)$ is

$$R = \iint P(x, y) \log \frac{P(x, y)}{P(x)P(y)} \, dx \, dy.$$

We define the rate R_1 of generating information for a given quality v_1 of reproduction to be the minimum of R when we keep v fixed at v_1 and vary $P_x(y)$. That is:

$$R_1 = \operatorname*{Min}_{P_x(y)} \iint P(x, y) \log \frac{P(x, y)}{P(x)P(y)} \, dx \, dy$$

subject to the constraint:

$$v_1 = \iint P(x, y)\rho(x, y) \, dx \, dy.$$

This means that we consider, in effect, all the communication systems that might be used and that transmit with the required fidelity. The rate of transmission in bits per second is calculated for each one and we choose that having the least rate. This latter rate is the rate we assign the source for the fidelity in question.

The justification of this definition lies in the following result:

Theorem 21: If a source has a rate R_1 for a valuation v_1 it is possible to encode the output of the source and transmit it over a channel of capacity C with fidelity as near v_1 as desired provided $R_1 \leq C$. This is not possible if $R_1 > C$.

The last statement in the theorem follows immediately from the definition of R_1 and previous results. If it were not true we could transmit more than C bits per second over a channel of capacity C. The first part of the theorem is proved by a method analogous to that used for Theorem 11. We may, in the first place, divide the (x, y) space into a large number of small cells and represent the situation as a discrete case. This will not change the evaluation function by more than an arbitrarily small amount (when the cells are very small) because of the continuity assumed for $\rho(x, y)$. Suppose that $P_1(x, y)$ is the particular system which minimizes the rate and gives R_1. We choose from the high probability y's a set at random containing

$$2^{(R_1+\epsilon)T}$$

members where $\epsilon \to 0$ as $T \to \infty$. With large T each chosen point will be connected by high probability lines (as in Fig. 10) to a set of x's. A calculation similar to that used in proving Theorem 11 shows that with large T almost all x's are covered by the fans from the chosen y points for almost all choices of the y's. The communication system to be used operates as follows: The selected points are assigned binary numbers. When a message x is originated it will (with probability approaching 1 as $T \to \infty$) lie within at least one of the fans. The corresponding binary number is transmitted (or one of them chosen arbitrarily if there are several) over the channel by suitable coding means to give a small probability of error. Since $R_1 \leq C$ this is possible. At the receiving point the corresponding y is reconstructed and used as the recovered message.

The evaluation v_1' for this system can be made arbitrarily close to v_1 by taking T sufficiently large. This is due to the fact that for each long sample of message $x(t)$ and recovered message $y(t)$ the evaluation approaches v_1 (with probability 1).

It is interesting to note that, in this system, the noise in the recovered message is actually produced by a kind of general quantizing at the transmitter and is not produced by the noise in the channel. It is more or less analogous to the quantizing noise in PCM.

29. The Calculation of Rates

The definition of the rate is similar in many respects to the definition of channel capacity. In the former

$$R = \operatorname*{Min}_{P_x(y)} \int\int P(x,y) \log \frac{P(x,y)}{P(x)P(y)} \, dx \, dy$$

with $P(x)$ and $v_1 = \int\int P(x,y)\rho(x,y) \, dx \, dy$ fixed. In the latter

$$C = \operatorname*{Max}_{P(x)} \int\int P(x,y) \log \frac{P(x,y)}{P(x)P(y)} \, dx \, dy$$

with $P_x(y)$ fixed and possibly one or more other constraints (e.g., an average power limitation) of the form $K = \int\int P(x,y) \lambda(x,y) \, dx \, dy$.

A partial solution of the general maximizing problem for determining the rate of a source can be given. Using Lagrange's method we consider

$$\iint \left[P(x, y) \log \frac{P(x, y)}{P(x)P(y)} + \mu\, P(x, y)\rho(x, y) \right.$$
$$\left. + \nu(x)P(x, y) \right] dx\, dy.$$

The variational equation (when we take the first variation on $P(x, y)$) leads to

$$P_y(x) = B(x)\, e^{-\lambda\rho(x,y)}$$

where λ is determined to give the required fidelity and $B(x)$ is chosen to satisfy

$$\int B(x)\, e^{-\lambda\rho(x,y)}\, dx = 1.$$

This shows that, with best encoding, the conditional probability of a certain cause for various received y, $P_y(x)$ will decline exponentially with the distance function $\rho(x, y)$ between the x and y in question.

In the special case where the distance function $\rho(x, y)$ depends only on the (vector) difference between x and y,

$$\rho(x, y) = \rho(x - y)$$

we have

$$\int B(x)\, e^{-\lambda\rho(x-y)}\, dx = 1.$$

Hence $B(x)$ is constant, say α, and

$$P_y(x) = \alpha e^{-\lambda\rho(x-y)}.$$

Unfortunately these formal solutions are difficult to evaluate in particular cases and seem to be of little value. In fact, the actual calculation of rates has been carried out in only a few very simple cases.

If the distance function $\rho(x, y)$ is the mean square discrepancy between x and y and the message ensemble is white noise, the rate can be determined. In that case we have

$$R = \text{Min}\ [H(x) - H_y(x)] = H(x) - \text{Max}\ H_y(x)$$

with $N = \overline{(x - y)^2}$. But the Max $H_y(x)$ occurs when $y - x$ is a white noise, and is equal to $W_1 \log 2\pi eN$ where W_1 is the band-width of the message ensemble. Therefore

$$R = W_1 \log 2\pi eQ - W_1 \log 2\pi eN$$
$$= W_1 \log \frac{Q}{N}$$

where Q is the average message power. This proves the following:

Theorem 22: The rate for a white noise source of power Q and band W_1 relative to an R.M.S. measure of fidelity is

$$R = W_1 \log \frac{Q}{N}$$

where N is the allowed mean square error between original and recovered messages.

More generally with any message source we can obtain inequalities bounding the rate relative to a mean square error criterion.

Theorem 23: The rate for any source of band W_1 is bounded by

$$W_1 \log \frac{Q_1}{N} \leq R \leq W_1 \log \frac{Q}{N}$$

where Q is the average power of the source, Q_1 its entropy power and N the allowed mean square error.

The lower bound follows from the fact that the Max $H_y(x)$ for a given $\overline{(x - y)^2} = N$ occurs in the white noise case. The upper bound results if we place the points (used in the proof of Theorem 21) not in the best way but at random in a sphere of radius $\sqrt{Q - N}$.

Acknowledgments

The writer is indebted to his colleagues at the Laboratories, particularly to Dr. H. W. Bode, Dr. J. R. Pierce, Dr. B. McMillan, and Dr. B. M. Oliver for many helpful suggestions and criticisms during the course of this work. Credit should also be given to Professor N. Wiener, whose elegant solution of the problems of filtering and prediction of stationary ensembles has considerably influenced the writer's thinking in this field.

Appendix 1.   **The Growth of the Number of Blocks of Symbols with a Finite State Condition**

Let $N_i(L)$ be the number of blocks of symbols of length L ending in state i. Then we have

$$N_j(L) = \sum_{i,s} N_i(L - b_{ij}^{(s)})$$

where $b_{ij}^1, b_{ij}^2, \cdots, b_{ij}^m$ are the length of the symbols which may be chosen in state i and lead to state j. These are linear difference equations and the behavior as $L \to \infty$ must be of the type

$$N_j = A_j W^L.$$

Substituting in the difference equation

$$A_j W^L = \sum_{i,s} A_i W^{L-b_{ij}^{(s)}}$$

or

$$A_j = \sum_{i,s} A_i W^{-b_{ij}^{(s)}}$$

$$\sum_i \left(\sum_s W^{-b_{ij}^{(s)}} - \delta_{ij} \right) A_i = 0.$$

For this to be possible the determinant

$$D(W) = |a_{ij}| = \left| \sum_s W^{-b_{ij}^{(s)}} - \delta_{ij} \right|$$

must vanish and this determines W, which is, of course, the largest real root of $D = 0$.

The quantity C is then given by

$$C = \lim_{L \to \infty} \frac{\log \Sigma A_j W^L}{L} = \log W$$

and we also note that the same growth properties result if we require that all blocks start in the same (arbitrarily chosen) state.

Appendix 2.   **Derivation of $H = -\Sigma\, p_i \log p_i$**

Let $H\left(\dfrac{1}{n}, \dfrac{1}{n}, \cdots, \dfrac{1}{n}\right) = A(n)$. From condition (3) we can decompose a choice from s^m equally likely possibilities into a series of m choices each from s equally likely possibilities and obtain

$$A(s^m) = m\,A(s).$$

Similarly
$$A(t^n) = n A(t).$$

We can choose n arbitrarily large and find an m to satisfy
$$s^m \leq t^n < s^{(m+1)}.$$

Thus, taking logarithms and dividing by $n \log s$,

$$\frac{m}{n} \leq \frac{\log t}{\log s} \leq \frac{m}{n} + \frac{1}{n} \quad \text{or} \quad \left| \frac{m}{n} - \frac{\log t}{\log s} \right| < \epsilon$$

where ϵ is arbitrarily small. Now from the monotonic property of $A(n)$,

$$A(s^m) \leq A(t^n) \leq A(s^{m+1})$$
$$m A(s) \leq nA(t) \leq (m + 1) A(s).$$

Hence, dividing by $nA(s)$,

$$\frac{m}{n} \leq \frac{A(t)}{A(s)} \leq \frac{m}{n} + \frac{1}{n} \quad \text{or} \quad \left| \frac{m}{n} - \frac{A(t)}{A(s)} \right| < \epsilon$$

$$\left| \frac{A(t)}{A(s)} - \frac{\log t}{\log s} \right| \leq 2\epsilon \qquad A(t) = K \log t$$

where K must be positive to satisfy (2).

Now suppose we have a choice from n possibilities with commeasurable probabilities $p_i = \frac{n_i}{\Sigma n_i}$ where the n_i are integers. We can break down a choice from Σn_i possibilities into a choice from n possibilities with probabilities $p_1, \cdots, p_n$ and then, if the ith was chosen, a choice from n_i with equal probabilities. Using condition (3) again, we equate the total choice from Σn_i as computed by two methods

$$K \log \Sigma n_i = H(p_1, \cdots, p_n) + K\Sigma \, p_i \log n_i.$$

Hence

$$H = K[\Sigma p_i \log \Sigma n_i - \Sigma p_i \log n_i]$$
$$= - K\Sigma p_i \log \frac{n_i}{\Sigma n_i} = - K\Sigma p_i \log p_i.$$

If the p_i are incommeasurable, they may be approximated by rationals and the same expression must hold by our continuity assumption. Thus the expression holds in general. The choice of coefficient K is a matter of convenience and amounts to the choice of a unit of measure.

Appendix 3. **Theorems on Ergodic Sources**

We assume the source to be ergodic so that the strong law of large numbers can be applied. Thus the number of times a given path p_{ij} in the network is traversed in a long sequence of length N is about proportional to the probability of being at i, say P_i, and then choosing this path, $P_i p_{ij} N$. If N is large enough the probability of percentage error $\pm \delta$ in this is less than ϵ so that for all but a set of small probability the actual numbers lie within the limits

$$(P_i p_{ij} \pm \delta) N.$$

Hence nearly all sequences have a probability p given by

$$p = \Pi p_{ij}^{(P_i p_{ij} \pm \delta) N}$$

and $\dfrac{\log p}{N}$ is limited by

$$\frac{\log p}{N} = \Sigma (P_i p_{ij} \pm \delta) \log p_{ij}$$

or

$$\left| \frac{\log p}{N} - \Sigma P_i p_{ij} \log p_{ij} \right| < \eta.$$

This proves Theorem 3.

Theorem 4 follows immediately from this on calculating upper and lower bounds for $n(q)$ based on the possible range of values of p in Theorem 3.

In the mixed (not ergodic) case if

$$L = \Sigma\ p_i L_i$$

and the entropies of the components are $H_1 \geq H_2 \geq \cdots \geq H_n$ we have the

Theorem: $\underset{N \to \infty}{\text{Lim}} \dfrac{\log n\ (q)}{N} = \varphi(q)$ *is a decreasing step function,*

$$\varphi(q) = H_s \text{ in the interval} \sum_1^{s-1} \alpha_i < q < \sum_1^s \alpha_i.$$

To prove Theorems 5 and 6 first note that F_N is monotonic decreasing because increasing N adds a subscript to a conditional entropy. A simple substitution for $p_{B_i}(S_j)$ in the definition of F_N shows that

$$F_N = N\, G_N - (N-1)\, G_{N-1}$$

and summing this for all N gives $G_N = \dfrac{1}{N}\, \Sigma\, F_N$. Hence $G_N \geq F_N$ and G_N monotonic decreasing. Also they must approach the same limit. By using Theorem 3 we see that $\underset{N \to \infty}{\mathrm{Lim}}\, G_N = H$.

Appendix 4.   Maximizing the Rate for a System of Constraints

Suppose we have a set of constraints on sequences of symbols that is of the finite state type and can be represented therefore by a linear graph, as in Fig. 2. Let $l_{ij}^{(s)}$ be the lengths of the various symbols that can occur in passing from state i to state j. What distribution of probabilities P_i for the different states and $p_{ij}^{(s)}$ for choosing symbol s in state i and going to state j maximizes the rate of generating information under these constraints? The constraints define a discrete channel and the maximum rate must be less than or equal to the capacity C of this channel, since if all blocks of large length were equally likely, this rate would result, and if possible this would be best. We will show that this rate can be achieved by proper choice of the P_i and $p_{ij}^{(s)}$. The rate in question is

$$\frac{-\sum\limits_{i,j,s} P_i p_{ij}^{(s)} \log p_{ij}^{(s)}}{\sum\limits_{i,j,s} P_i p_{ij}^{(s)} l_{ij}^{(s)}}.$$

Let

$$p_{ij}^{(s)} = \frac{B_j}{B_i}\, W^{-l_{ij}^{(s)}}$$

where the B_i satisfy the equations

$$B_i = \sum_{j,s} B_j W^{-l_{ij}^{(s)}}.$$

This homogeneous system has a non-vanishing solution since W is such that the determinant of the coefficients is zero:

$$\left| \sum_s W^{-l_{ij}^{(s)}} - \delta_{ij} \right| = 0.$$

The $p_{ij}^{(s)}$ defined thus are satisfactory transition probabilities for in the first place,

$$\sum_{j,s} p_{ij}^{(s)} = \sum_{j,s} \frac{B_j}{B_i} W^{-l_{ij}^{(s)}}$$

$$= \frac{B_i}{B_i} = 1$$

so that the sum of the probabilities from any particular junction point is unity. Furthermore they are non-negative as can be seen from a consideration of the quantities A_i given in Appendix 1. The A_i are necessarily non-negative and the B_i satisfy a similar system of equations but with i and j interchanged. This amounts to reversing the orientation on the lines of the graph.

Substituting the assumed values of $p_{ij}^{(s)}$ in the general equation for the rate we obtain

$$- \frac{\Sigma P_i p_{ij}^{(s)} \log \frac{B_j}{B_i} W^{-l_{ij}^{(s)}}}{\Sigma P_i p_{ij}^{(s)} l_{ij}^{(s)}}$$

$$= \frac{\log W \Sigma P_i p_{ij}^{(s)} l_{ij}^{(s)} - \Sigma P_i p_{ij}^{(s)} \log B_j + \Sigma P_i p_{ij}^{(s)} \log B_i}{\Sigma P_i p_{ij}^{(s)} l_{ij}^{(s)}}$$

$$= \log W = C.$$

Hence the rate with this set of transition probabilities is C and since this rate could never be exceeded this is the maximum.

Appendix 5

Let S_1 be any measurable subset of the g ensemble, and S_2 the subset of the f ensemble which gives S_1 under the operation T. Then

$$S_1 = TS_2.$$

Let H^λ be the operator which shifts all functions in a set by the time λ. Then

$$H^\lambda S_1 = H^\lambda TS_2 = TH^\lambda S_2$$

since T is invariant and therefore commutes with H^λ. Hence if $m[S]$ is the probability measure of the set S

$$m[H^\lambda S_1] = m[TH^\lambda S_2] = m[H^\lambda S_2]$$
$$= m[S_2] = m[S_1]$$

where the second equality is by definition of measure in the g space, the third since the f ensemble is stationary, and the last by definition of g measure again. This shows that the g ensemble is stationary.

To prove that the ergodic property is preserved under invariant operations, let S_1 be a subset of the g ensemble which is invariant under H^λ, and let S_2 be the set of all functions f which transform into S_1. Then

$$H^\lambda S_1 = H^\lambda T S_2 = T H^\lambda S_2 = S_1$$

so that $H^\lambda S_2$ is included in S_2 for all λ. Now, since

$$m[H^\lambda S_2] = m[S_2] = m[S_1]$$

this implies

$$H^\lambda S_2 = S_2$$

for all λ with $m[S_2] \neq 0, 1$. This contradiction shows that S_1 does not exist.

Appendix 6

The upper bound, $\overline{N}_3 \leq N_1 + N_2$, is due to the fact that the maximum possible entropy for a power $N_1 + N_2$ occurs when we have a white noise of this power. In this case the entropy power is $N_1 + N_2$.

To obtain the lower bound, suppose we have two distributions in n dimensions $p(x_i)$ and $q(x_i)$ with entropy powers $\overline{N}_1$ and $\overline{N}_2$. What form should p and q have to minimize the entropy power $\overline{N}_3$ of their convolution $r(x_i)$:

$$r(x_i) = \int p(y_i) q(x_i - y_i) \, dy_i.$$

The entropy H_3 of r is given by

$$H_3 = -\int r(x_i) \log r(x_i) \, dx_i.$$

We wish to minimize this subject to the constraints

$$H_1 = - \int p(x_i) \log p(x_i) \, dx_i$$

$$H_2 = - \int q(x_i) \log q(x_i) \, dx_i.$$

We consider then

$$U = - \int [r(x) \log r(x) + \lambda p(x) \log p(x) + \mu q(x) \log q(x)] \, dx$$

$$\delta U = - \int [[1 + \log r(x)] \, \delta r(x) + \lambda [1 + \log p(x)] \, \delta p(x) \\ + \mu [1 + \log q(x)] \, \delta q(x)] \, dx.$$

If $p(x)$ is varied at a particular argument $x_i = s_i$, the variation in $r(x)$ is

$$\delta r(x) = q(x_i - s_i)$$

and

$$\delta U = - \int q(x_i - s_i) \log r(x_i) \, dx_i - \lambda \log p(s_i) = 0$$

and similarly when q is varied. Hence the conditions for a minimum are

$$\int q(x_i - s_i) \log r(x_i) \, dx_i = - \lambda \log p(s_i)$$

$$\int p(x_i - s_i) \log r(x_i) \, dx_i = - \mu \log q(s_i).$$

If we multiply the first by $p(s_i)$ and the second by $q(s_i)$ and integrate with respect to s_i we obtain

$$H_3 = - \lambda H_1$$

$$H_3 = - \mu H_2$$

or solving for λ and μ and replacing in the equations

$$H_1 \int q(x_i - s_i) \log r(x_i) \, dx_i = - H_3 \log p(s_i)$$

$$H_2 \int p(x_i - s_i) \log r(x_i) \, dx_i = - H_3 \log q(s_i).$$

Now suppose $p(x_i)$ and $q(x_i)$ are normal

$$p(x_i) = \frac{|A_{ij}|^{n/2}}{(2\pi)^{n/2}} \exp - \tfrac{1}{2} \Sigma A_{ij} x_i x_j$$

$$q(x_i) = \frac{|B_{ij}|^{n/2}}{(2\pi)^{n/2}} \exp - \tfrac{1}{2} \Sigma B_{ij} x_i x_j.$$

Then $r(x_i)$ will also be normal with quadratic form C_{ij}. If the inverses of these forms are a_{ij}, b_{ij}, c_{ij} then

$$c_{ij} = a_{ij} + b_{ij}.$$

We wish to show that these functions satisfy the minimizing conditions if and only if $a_{ij} = Kb_{ij}$ and thus give the minimum H_3 under the constraints. First we have

$$\log r(x_i) = \frac{n}{2} \log \frac{1}{2\pi} \mid C_{ij} \mid - \tfrac{1}{2} \Sigma C_{ij} x_i x_j$$

$$\int q(x_i - s_i) \log r(x_i) = \frac{n}{2} \log \frac{1}{2\pi} \mid C_{ij} \mid - \tfrac{1}{2} \Sigma C_{ij} s_i s_j - \tfrac{1}{2} \Sigma C_{ij} b_{ij}.$$

This should equal

$$\frac{H_3}{H_1} \left[\frac{n}{2} \log \frac{1}{2\pi} \mid A_{ij} \mid - \tfrac{1}{2} \Sigma A_{ij} s_i s_j \right]$$

which requires $A_{ij} = \dfrac{H_1}{H_3} C_{ij}.$

In this case $A_{ij} = \dfrac{H_1}{H_2} B_{ij}$ and both equations reduce to identities.

Appendix 7

The following will indicate a more general and more rigorous approach to the central definitions of communication theory. Consider a probability measure space whose elements are ordered pairs (x, y). The variables x, y are to be identified as the possible transmitted and received signals of some long duration T. Let us call the set of all points whose x belongs to a subset S_1 of x points the strip over S_1, and similarly the set whose y belong to S_2 the strip over S_2. We divide x and y into a collection of non-overlapping measurable subsets X_i and Y_i approximate to the rate of transmission R by

$$R_1 = \frac{1}{T} \sum_i P(X_i, Y_i) \log \frac{P(X_i, Y_i)}{P(X_i)P(Y_i)}$$

where

$P(X_i)$ is the probability measure of the strip over X_i

$P(Y_i)$ is the probability measure of the strip over Y_i

$P(X_i, Y_i)$ is the probability measure of the intersection of the strips.

A further subdivision can never decrease R_1. For let X_1 be divided into $X_1 = X_1' + X_1''$ and let

$$P(Y_1) = a \qquad\qquad P(X_1) = b + c$$
$$P(X_1') = b \qquad\qquad P(X_1', Y_1) = d$$
$$P(X_1'') = c \qquad\qquad P(X_1'', Y_1) = e$$
$$P(X_1, Y_1) = d + e.$$

Then in the sum we have replaced (for the X_1, Y_1 intersection)

$$(d + e) \log \frac{d + e}{a(b + c)} \text{ by } d \log \frac{d}{ab} + e \log \frac{e}{ac}.$$

It is easily shown that with the limitation we have on $b, c, d, e,$

$$\left[\frac{d + e}{b + c}\right]^{d+e} \leq \frac{d^d \cdot e^e}{b^d\, c^e}$$

and consequently the sum is increased. Thus the various possible subdivisions form a directed set, with R monotonic increasing with refinement of the subdivision. We may define R unambiguously as the least upper bound for the R_1 and write it

$$R = \frac{1}{T} \iint P(x, y) \log \frac{P(x, y)}{P(x)P(y)}\, dx\, dy.$$

This integral, understood in the above sense, includes both the continuous and discrete cases and of course many others which cannot be represented in either form. It is trivial in this formulation that if x and u are in one-to-one correspondence, the rate from u to y is equal to that from x to y. If v is any function of y (not necessarily with an inverse) then the rate from x to y is greater than or equal to that from x to v since, in the calculation of the approximations, the subdivisions of y are essentially a finer subdivision of those for v. More generally if y and v are related not functionally but statistically, i.e., we have a probability measure space (y,v), then $R(x,v) \leq R(x,y)$. This means that any operation applied to the received signal, even though it involves statistical elements, does not increase R.

Another notion which should be defined precisely in an abstract formulation of the theory is that of "dimension rate," that is the average number of dimensions required per second to specify a member of an ensemble. In the band limited case $2W$ numbers per second are sufficient. A general definition can be framed as follows. Let $f_\alpha(t)$ be an ensemble of functions and let $\rho_T[f_\alpha(t), f_\beta(t)]$ be a metric measuring the "distance" from f_α to f_β over the time T (for example the R.M.S. discrepancy over this interval). Let $N(\epsilon, \delta, T)$ be the least number of elements f which can be chosen such that all elements of the ensemble apart from a set of measure δ are within the distance ϵ of at least one of those chosen. Thus we are covering the space to within ϵ apart from a set of small measure δ. We define the dimension rate λ for the ensemble by the triple limit

$$\lambda = \lim_{\delta \to 0} \lim_{\epsilon \to 0} \lim_{T \to \infty} \frac{\log N(\epsilon, \delta, T)}{T \log \epsilon}.$$

This is a generalization of the measure type definitions of dimension in topology, and agrees with the intuitive dimension rate for simple ensembles where the desired result is obvious.

Claude E. Shannon is retired from his position as research mathematician at the Bell Telephone Laboratories. From 1958 to 1978 he was also Donner Professor of Science at the Massachusetts Institute of Technology.

Warren Weaver, now deceased, had a distinguished career in academic, government, and foundation work.

Richard E. Blahut and Bruce Hajek are professors of electrical and computer engineering at the University of Illinois at Urbana-Champaign.

.

UNIVERSITY OF ILLINOIS PRESS
1325 SOUTH OAK STREET
CHAMPAIGN, ILLINOIS 61820-6903
WWW.PRESS.UILLINOIS.EDU